BIOLOGY
THROUGH A MICROSCOPE

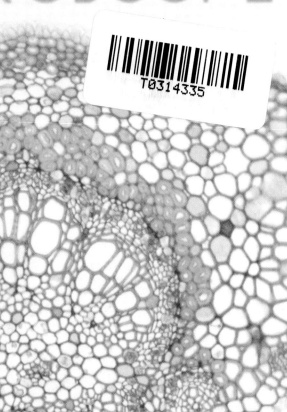

Ficus (dicot)
leaf, c.s.

MasterBooks®
CURRICULUM

Author: Chris Hallski

Master Books Creative Team:

Editor: Laura Welch

Design: Diana Bogardus

Cover Design: Diana Bogardus

Copy Editors:
Judy Lewis
Willow Meek
Craig Froman

Curriculum Review:
Kristen Pratt
Laura Welch
Diana Bogardus

First printing: October 2019
Second printing: March 2023

Master Books, P.O. Box 726, Green Forest, AR 72638

Master Books® is a division of the New Leaf Publishing Group, Inc.

ISBN: 978-1-68344-191-5
ISBN: 978-1-61458-731-6 (digital)

Library of Congress: 2019950592

Printed in the United States of America

Please visit our website for other great titles:
www.masterbooks.com

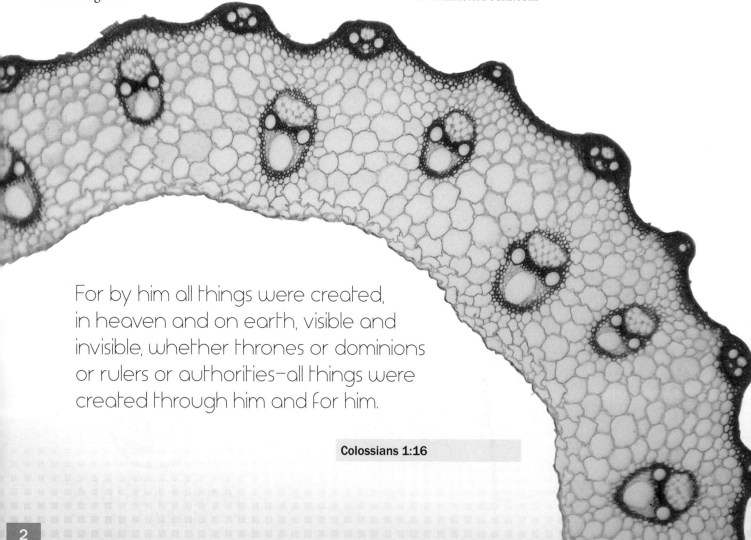

For by him all things were created, in heaven and on earth, visible and invisible, whether thrones or dominions or rulers or authorities—all things were created through him and for him.

Colossians 1:16

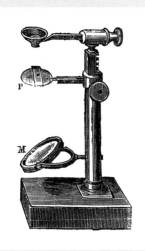

Antique microscopes

For thousands of years, people have marveled at the world God created, but it wasn't until the invention of the microscope in the 13th century that we would finally see the how detailed and beautiful His unseen creations were.

The first forms of microscope were simple magnifying glasses and eyeglasses. Early compound microscopes are the first that we would recognize as a microscope today, with an objective lens near the specimen being viewed and a separate eyepiece lens. Simple mirrors and separate light sources provided illumination.

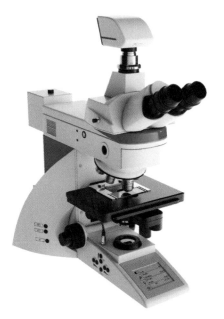

By the 20th century, other methods of illuminating the specimen were developed. Köhler illumination is particularly useful because it results in an even illumination of the sample with no visible details of the light source. Advances in optics also allowed for binocular microscopes with two eyepieces and the attachment of cameras.

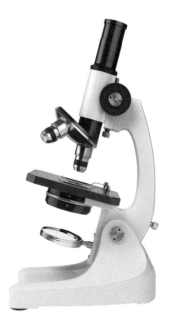

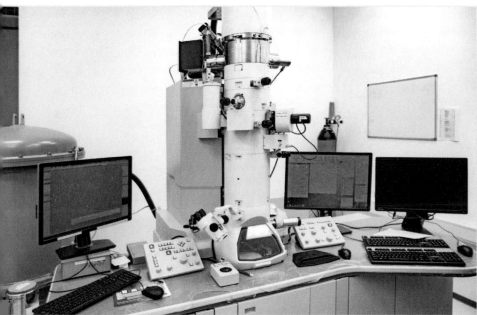

Currently, specialized microscopes use electrons, scanning probes, fluorescence, and x-rays to produce images far beyond what can normally be seen. Computer colorization of the images helps scientists see fine details as they study them.

In spite of all the advancements throughout the years, light microscopes are still similar to the original microscopes with light sources, objective lenses, and eyepieces. Most basic microscopes today will share many common parts, as shown below:

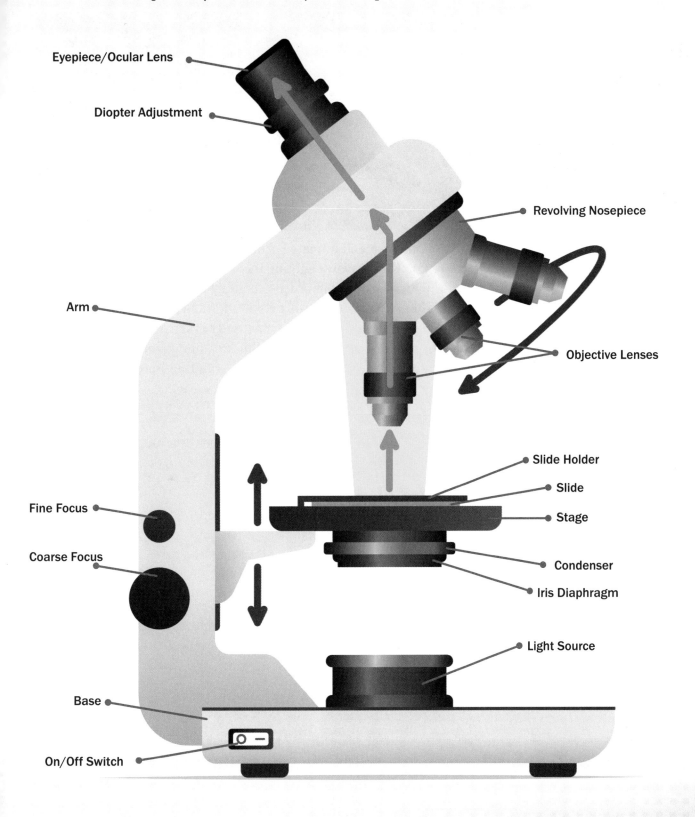

Eyepiece/Ocular Lens

Diopter Adjustment

Arm

Fine Focus

Coarse Focus

Base

On/Off Switch

Revolving Nosepiece

Objective Lenses

Slide Holder

Slide

Stage

Condenser

Iris Diaphragm

Light Source

Even simple light microscopes have many available options for illuminating specimens depending on the sample. The examples below were taken of the same slide of colored threads at a low magnification (40x).

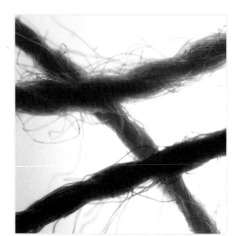 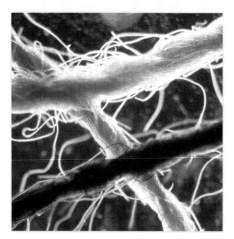 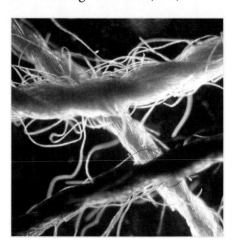

This sample is using a typical brightfield backlit setup with the lamp underneath the specimen and no special attachments. Thicker samples might be illuminated from the top, but most thin slices or small specimens are adequately lit using a light source under the microscope's stage.

These samples were illuminated using phase contrast and darkfield techniques, which involve lighting the sample from the edges with specific patterns of light in order to highlight details that are not clearly visible. These techniques require some additional attachments or filters on the condenser, but are vital for the study of samples with subtle details or minimal contrast against the containing **medium**. Marine diatoms are an excellent example of this, with transparent **cell walls** nearly indistinguishable from the water they live in.

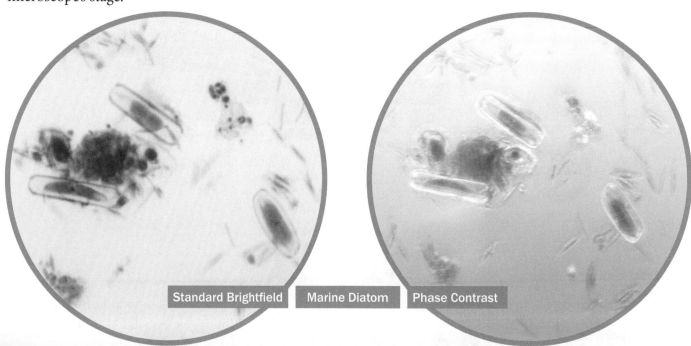

Standard Brightfield | Marine Diatom | Phase Contrast

The images of marine diatoms above illustrate the difference between brightfield and phase contrast microscopy. The samples are shown at equivalent sizes: the first image is a standard brightfield image, while the second is using phase contrast to highlight the details of the diatoms' cell walls.

Salt Crystals

Pink Himalayan Salt Crystals

Himalayan Rock Salt

Kala Namak (India)

Alaea Red Salt (Hawaii)

Gerand Salt

Herb Salt

Black Pearl Salt (Hawaii)

Himalayan Salt

Common Salt

Atlantic Sea Salt

The salts shown above are all comprised of sodium chloride with variations in added compounds, both naturally occurring and added during processing. The difference in coloration is easily visible without assistance, but a microscope is necessary to study differences in crystal structures caused by the rate of formation and analyze distinctions like those between salt crystals that appear white because of mineral deposits and ones that are full of tiny air bubbles.

Brown Zircon

Blue Zircon

These microscopy techniques are used for studying other crystals as well. Many minerals form crystalline structures, from gemstones to salts.

A microscope is used for studying a crystal's optical properties and understanding its growth patterns, locating defects and identifying impurities. These imperfections frequently result in color variations, like over 80 minerals and trace elements contributing to Pink Himalayan Salt's distinctive color.

In cases where the color is caused by damaged crystalline structure, it can sometimes be altered by heating the crystal or removed when faceting it. An example of this is when natural brown zircon is heat treated to alter its color and then cut to remove surface defects.

Faceted Blue Zircon

KINGDOM
SAMPLES
➡ ANIMALS

Animals have three different types of muscle: cardiac, smooth, and skeletal. Samples of these are shown below in various magnifications.

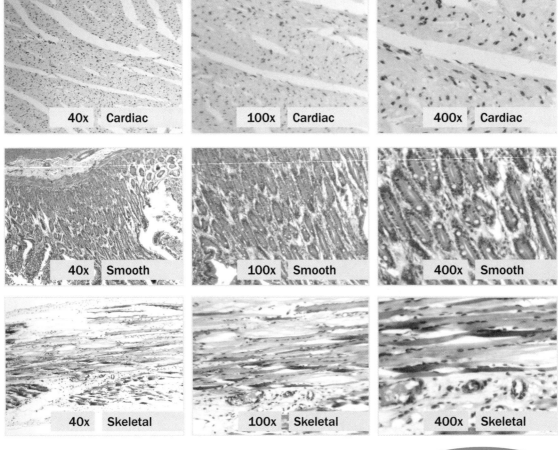

40x   Cardiac

100x   Cardiac

400x   Cardiac

Cardiac muscle is a **striated** involuntary muscle type found in the walls of the heart.

40x   Smooth

100x   Smooth

400x   Smooth

Smooth muscle fibers are also involuntary, and are found in the walls of internal organs other than the heart.

40x   Skeletal

100x   Skeletal

400x   Skeletal

Skeletal muscle fibers are striated voluntary muscles. They have banding similar to cardiac muscle visible in the individual fibers.

Muscles of all types receive signals from the brain via nerve cells called motor neurons. These are too small to be easily seen with without magnification, but are easily visible under a microscope. Motor neurons are divided into two groups: upper motor neurons that connect directly to the brain and transmit signals to the spinal cord and lower motor neurons that connect the spinal cord to individual muscle fibers. Both upper and lower motor neurons consist of several dendrites to receive signals from other neurons and propagate them into the cell body, and a single axon, the nerve fiber that carries the received signal through the body.

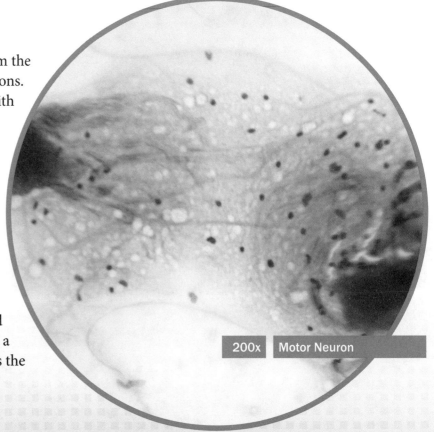

200x   Motor Neuron

Covering the muscles, nerves, and other internal organs is the integumentary system, comprising skin, hair, scales, feathers, and nails. This system is a body's first line of defense against infection and physical damage. While skin itself is easily visible on most animals, finer structural details can only be studied under magnification, such as the hair follicles below. Other structures like fish scales and individual hairs are also easier to study under magnification.

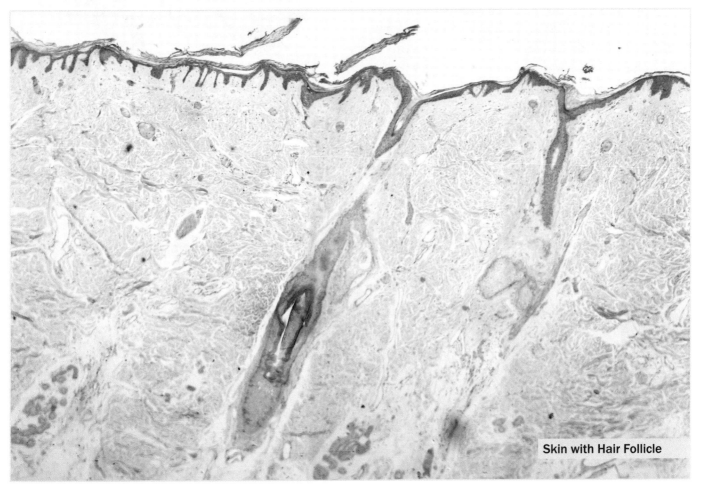

Skin with Hair Follicle

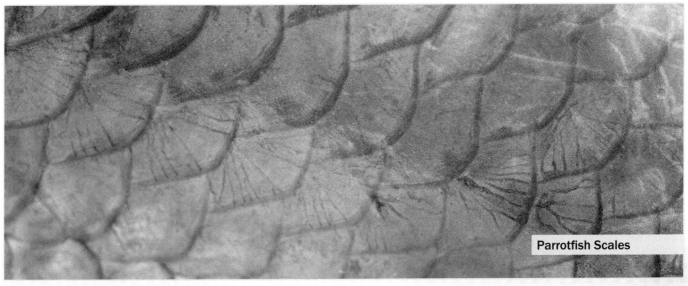

Parrotfish Scales

General Biology: Animal Hair / Blood

Hair is divided into two major structures: the follicle, or bulb, which is the portion of hair that grows, and the shaft, or visible strand that protrudes from the skin. Hair shafts are divided into zones: the cuticle, cortex, and in thicker hairs an unstructured central area known as the medulla. The cuticle encloses the hair with thin layers of cells overlapping like shingles and protects it from damage. The cortex contains keratin bundles that shape the type of hair and melanin pigments that determine its color. Rounder cortices form straighter hairs, while oval cross sections form wavier or curly hairs.

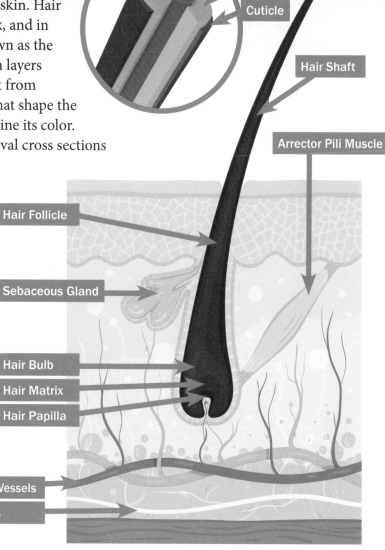

Medulla

Cortex

Cuticle

Hair Shaft

Arrector Pili Muscle

Hair Follicle

Sebaceous Gland

Hair Bulb

Hair Matrix

Hair Papilla

Blood Vessels

Nerves

Human Hairs with Follicles

125x   Porcupine Quill (SEM)

Hair provides many animals with protection from both UV radiation in sunlight and extremes of temperature. The arrector pili muscle is a tiny involuntary muscle that attaches to each hair follicle individually, causing the hair to rise up, increasing trapped air among the erect hairs, which improves the insulating ability of fur. In some animals, like porcupines, the spines are special hairs which also stand erect and form a defensive barrier.

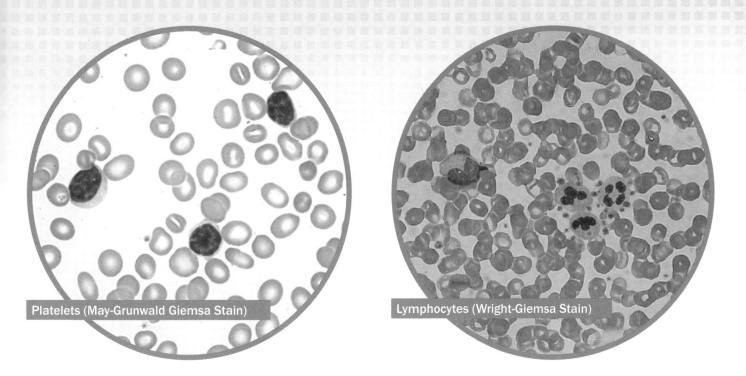

Platelets (May-Grunwald Giemsa Stain)

Lymphocytes (Wright-Giemsa Stain)

All members of the animal kingdom use blood for oxygen and nutrient transport throughout their bodies. One of the challenges of studying blood through a microscope is how to differentiate different cells or even organisms in the blood. An auxiliary technique for microscopic observation developed to help accomplish this is called staining.

Many different staining compounds exist with various behaviors, but all of them rely on the principle that certain cells or compounds will retain a specific dye better than other cells. This means that a staining compound that is absorbed by a **cell wall** will mark bacteria with a thick, exposed cell wall, but not bacteria with a thinner cell wall and outer membrane. This is referred to as Gram staining, and can identify whether a particular bacteria sample is likely to be generally antibiotic resistant.

Various Stained & Unstained Samples

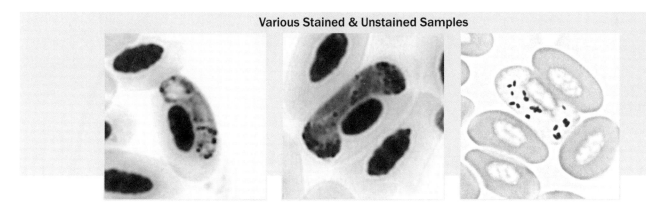

In blood samples, blends like a combination of the Wright and Giemsa stains are often used for differentiating cell types in a complete blood count, or the number of red and white blood cells, platelets, and their density in the blood. With this stain, red blood cells will appear pink, platelets a lighter pink and white blood cells will be various shades of blue. Other stains are used to identify foreign organisms in the blood such as bacteria and **parasites**. Some examples of these stains and the differences in how samples appear after staining are shown above.

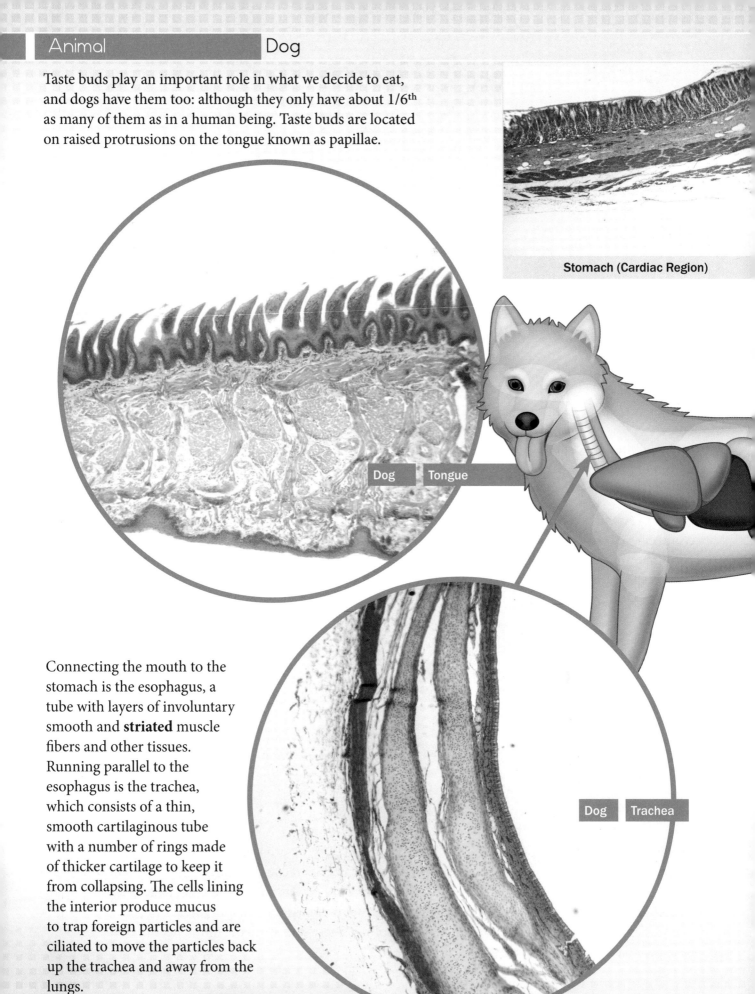

Taste buds play an important role in what we decide to eat, and dogs have them too: although they only have about 1/6th as many of them as in a human being. Taste buds are located on raised protrusions on the tongue known as papillae.

Stomach (Cardiac Region)

Dog | Tongue

Connecting the mouth to the stomach is the esophagus, a tube with layers of involuntary smooth and **striated** muscle fibers and other tissues. Running parallel to the esophagus is the trachea, which consists of a thin, smooth cartilaginous tube with a number of rings made of thicker cartilage to keep it from collapsing. The cells lining the interior produce mucus to trap foreign particles and are ciliated to move the particles back up the trachea and away from the lungs.

Dog | Trachea

The stomach is divided into three major areas with differing cell structures depending on their functions:

1. The cardiac region, which contains mucus secreting glands and is next to the esophagus.
2. The fundus region, where most of the mechanical action of breaking down food takes place.
3. The pyloric region, which secretes more mucus and the hormone gastrin, which aids digestion by increasing gastric contractions known as motility.

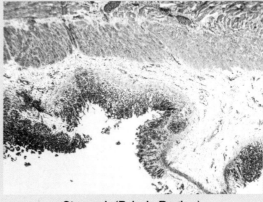

Stomach (Pyloric Region)

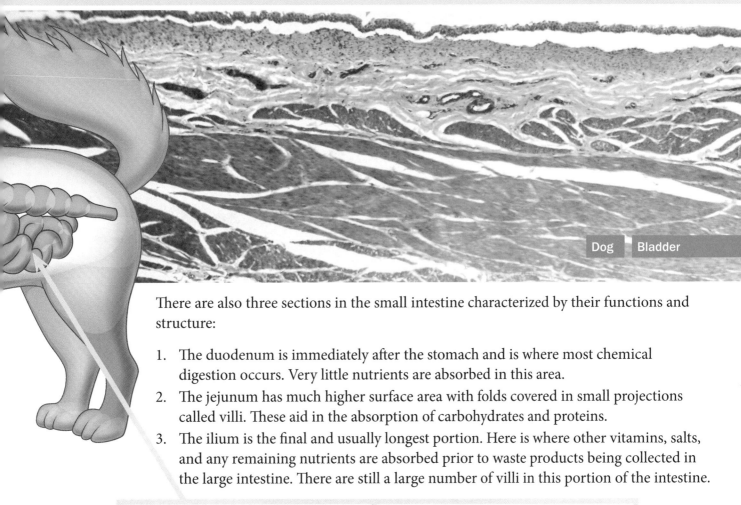

Dog | Bladder

There are also three sections in the small intestine characterized by their functions and structure:

1. The duodenum is immediately after the stomach and is where most chemical digestion occurs. Very little nutrients are absorbed in this area.
2. The jejunum has much higher surface area with folds covered in small projections called villi. These aid in the absorption of carbohydrates and proteins.
3. The ilium is the final and usually longest portion. Here is where other vitamins, salts, and any remaining nutrients are absorbed prior to waste products being collected in the large intestine. There are still a large number of villi in this portion of the intestine.

Duodenum **Jejunum** **Ilium**

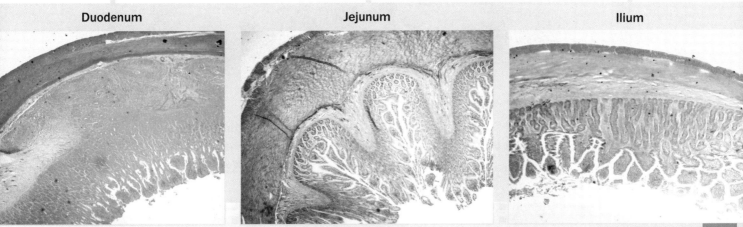

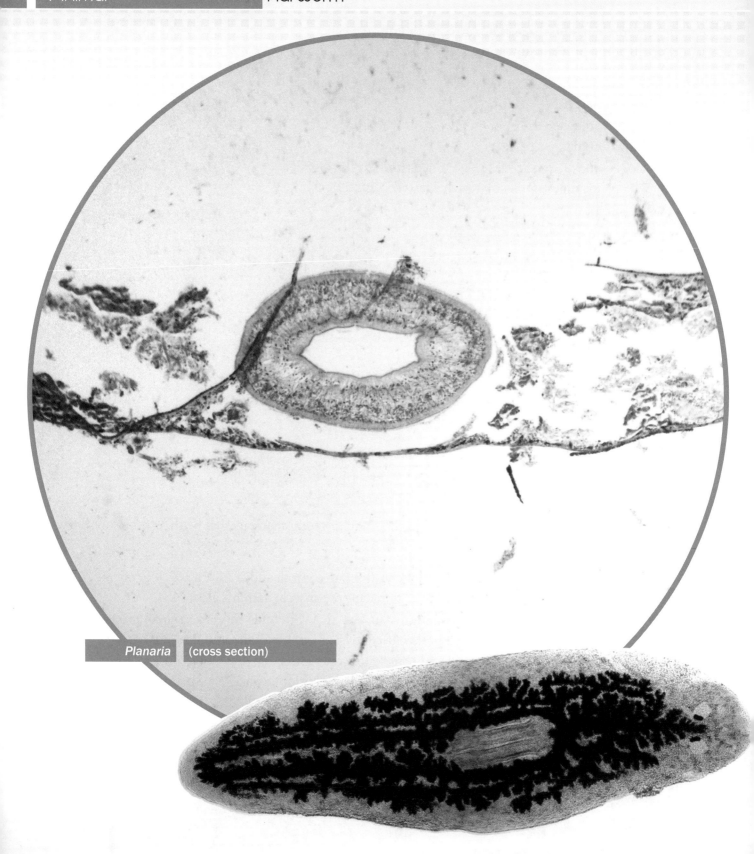

Planaria   (cross section)

The term **flatworm** encompasses a wide variety of simple, bilaterally symmetrical invertebrates. **Free-living** flatworms such as planaria do not require a host and are generally predatory: feeding on smaller organisms such as mosquito **larvae**.

Other species fall under classifications such as trematoda and cercomeromorpha, which are **parasitic** organisms more commonly known as flukes and tapeworms.

Flukes have a very complex life cycle, infecting multiple different hosts during their parasitic lifecycle. Beginning from an egg, the young fluke hatches into a larval form known as a **miracidium**. If the egg hasn't been eaten by a suitable host organism before hatching, the miracidium must quickly find a host or it will starve before entering the next phases of life. These phases depend on the species, but include a variety of intermediate forms such as **sporocysts**, **rediae**, and **cercariae** before transforming into the adult form.

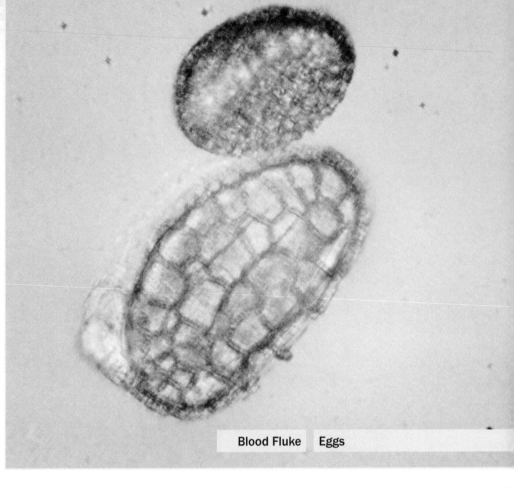

Blood Fluke   Eggs

Most trematodes require two hosts throughout their life cycle: typically the intermediate host is an invertebrate like a mollusc, and hosts the flatworm throughout the larval stages. The majority of adult forms are **hermaphroditic** and prefer vertebrate hosts where they reproduce sexually, although there are species of flatworms that grow to maturity in the initial host.

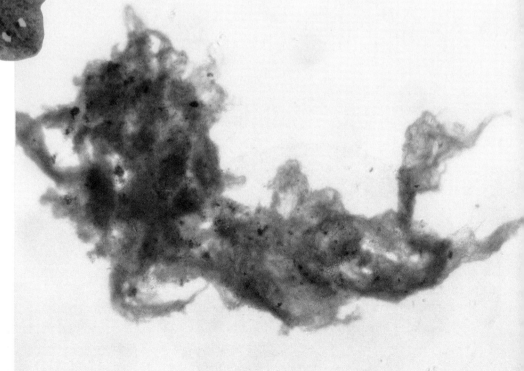

Trematoda   Miracidium

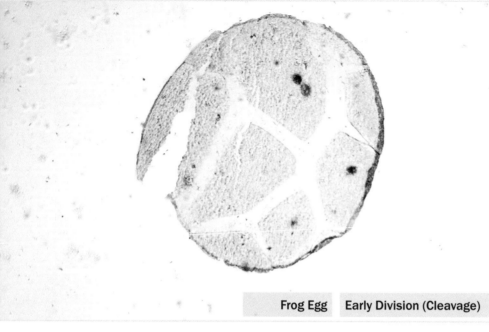

Frog Egg | Early Division (Cleavage)

Frog embryos start as a single-celled egg, but split into many smaller cells fairly rapidly through holoblastic cleavage (a process where the cell divides completely into smaller cells without increasing in mass). This process results in an embryo known as a morula, consisting of 16 to 64 cells.

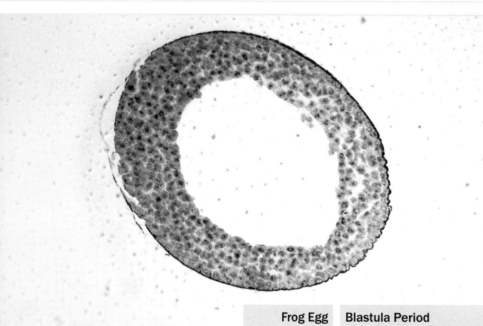

Frog Egg | Blastula Period

By the 128-cell stage, the cells have formed a sphere with a defined hollow, or blastocoel, in the middle. This form is considered a blastula, and separates the cells so they can differentiate into different types.

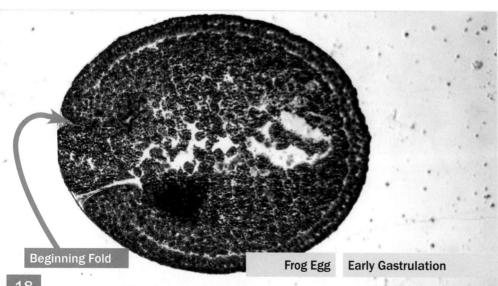

Beginning Fold

Frog Egg | Early Gastrulation

As the embryo continues to grow, the blastula folds in on itself in a process called gastrulation, forming several layers of cells in the process: the endoderm, ectoderm, and mesoderm. The endoderm will become the nervous system and skin, while the ectoderm forms muscle, connective tissue, and some internal organs. The mesoderm forms digestive organs, lungs, and other internal organs.

The development of the frog continues for 12 weeks, growing from a tadpole through the two and four-leg stages into a froglet with a stub of a tail. Finally, after a total of 16 weeks, the embryo has grown into a mature, tailless amphibian that is ready to reproduce.

Frog Skin | 1000x

Frog blood is very similar to most other vertebrates' blood: it still carries oxygen in the hemoglobin molecule and has a "filled-in-donut" shape. Other creatures have a variety of oxygen-carrying molecules such as haemocyanin, haemorythrin, and chlorocruorin that make their blood different colors. Green blood isn't limited to science fiction!

Frog Blood | Shown in Darkfield at 400x, and Brightfield at 400x & 1000x

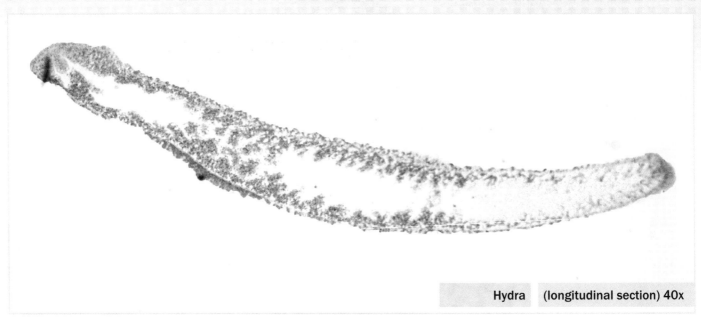

Hydra   (longitudinal section) 40x

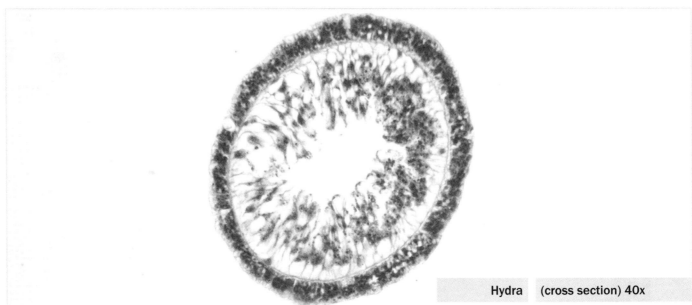

Hydra   (cross section) 40x

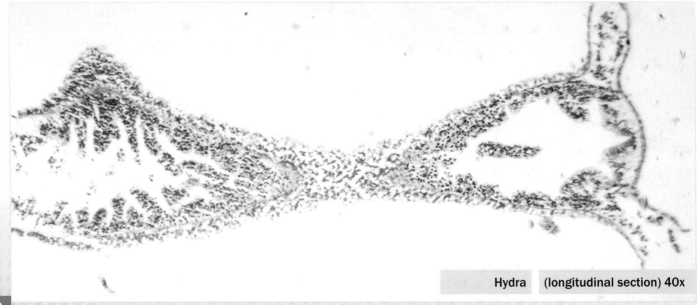

Hydra   (longitudinal section) 40x

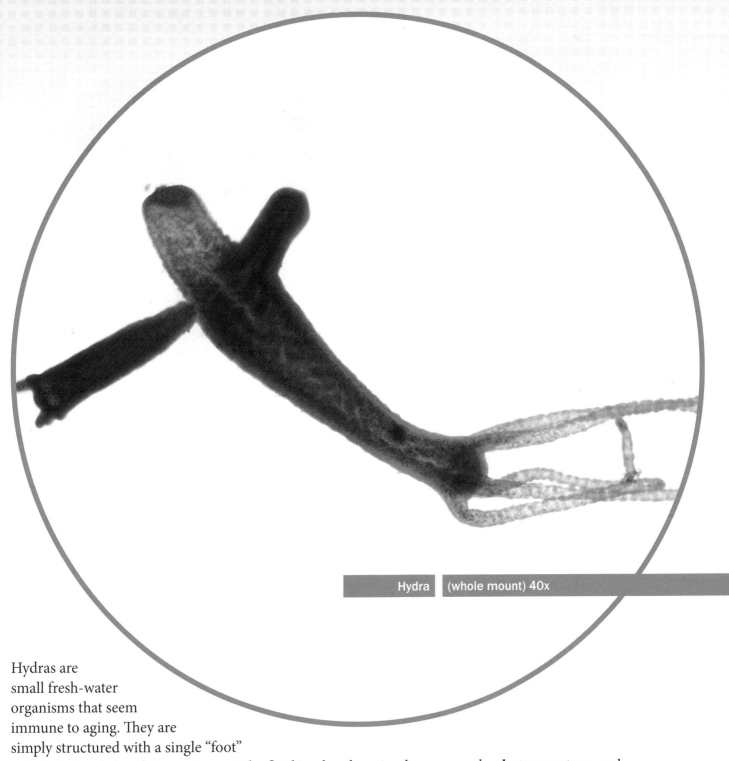

Hydra | (whole mount) 40x

Hydras are small fresh-water organisms that seem immune to aging. They are simply structured with a single "foot" called the basal disc that secretes a sticky fluid to glue them in place, a mouth, **photoreceptors**, and one to twelve tentacles. Their nervous system is a nerve net that enables reflexive reactions to stimuli and coordinated movement by looping and somersaulting: a process of bending over to grip the surface they are on with their mouth and tentacles, then continuing the motion lifting the foot over in a somersault to its new location. They can move several inches per day using this method.

Hydra hunt by extending their tentacles and waiting for prey to make contact. When a prey makes contact with a tentacle, harpoon-like cells known as nematocysts fire into the prey and the tentacles draw it into the hydra's mouth aperture to be enveloped. Indigestible material is discharged back out of the mouth after several days.

Mouse | Cuboidal Epithelium

Mice are tiny creatures, but the same types of tissues and organs that exist in mice are common to many mammals. This means that mice and rats are ideal test subjects for laboratory work, including genetics and behavioral studies. On the other hand, the small size of mouse organs means that a microscope is often necessary to study them closely.

On the next page, we are taking a closer look at a mouse kidney: shown full-size in cross section on a slide, as well as zoomed in. Kidneys in mice, like most mammals, perform a vital function of filtering excess liquid and waste products out of the bloodstream. The cells that perform these functions are known as epithelial cells which are described with two different classifications: by the number of cell layers and by the shape of the cells.

Simple epithelia are composed of a single layer of cells, while stratified epithelia consist of two or more layers. When describing the cell shapes, terms of squamous, cuboidal, and columnar are used to describe the cells. Squamous cells are extremely thin and flat; cuboidal cells are squarish in cross section; and columnar cells are tall and thin.

These descriptions of layering and cell shape are combined to define any particular cellular tissue as one of six possible types:

1. Simple squamous epithelia
2. Simple cuboidal epithelia
3. Simple columnar epithelia
4. Stratified squamous epithelia
5. Stratified cuboidal epithelia
6. Stratified columnar epithelia

As if that wasn't confusing enough, there are also two subclassifications called pseudostratified and transitional epithelia: simple columnar epithelium with uneven widths that appears to have multiple layers and stratified epithelium that can expand and contract as needed.

Epithelial tissues perform several vital functions: including separating organs from each other in the body and providing a buffer to control rates of interchange. One key characteristic of epithelial tissue is that the junctions between cells are very tight, in essence a waterproofing layer for the body. By preventing the passage of molecules between cells, epithelial tissue allows organs like the kidney to filter fluids and maintain electrolyte balance in the body through controlling the absorption and release of different elements.

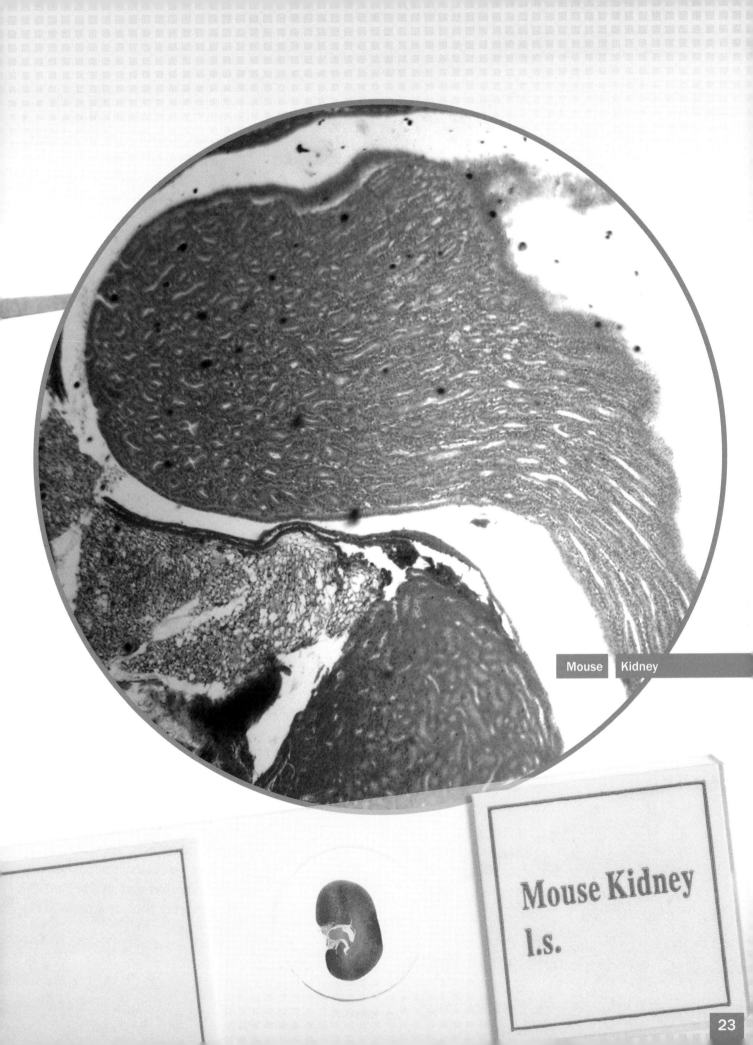

Mouse Kidney

Mouse Kidney
l.s.

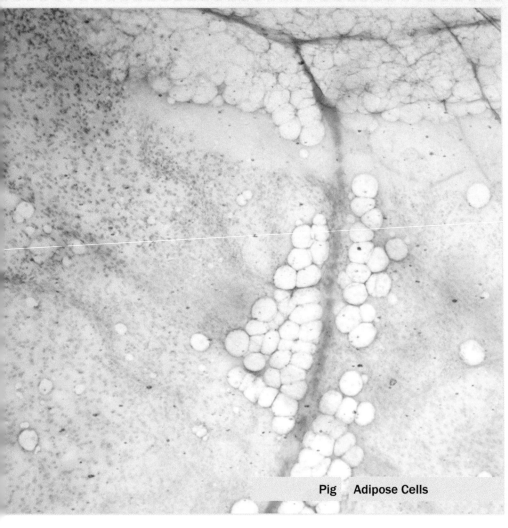

Pig   Adipose Cells

Fat cells, also known as adipose tissue, are an animal's primary method of storing energy for later use, carrying more than double the energy equivalent of proteins and carbohydrates per gram. While these cells are efficient at storing energy, animal diets don't always provide the exact nutrients needed when they are required, so some conversion is required. The liver is the predominate organ of the body when it comes to performing these functions of fat metabolism, both in converting excess carbohydrates and proteins into fatty acids and triglycerides for storage in adipose tissue as well as the production of bile necessary for digestion of ingested fats.

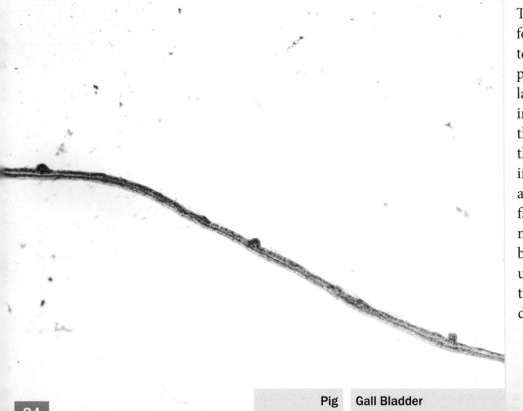

Pig   Gall Bladder

The gall bladder is not essential for survival, but it provides temporary storage for bile produced by the liver that is later released into the small intestine after eating to aid in the absorption of fats. Removing the gallbladder from a healthy individual may cause diarrhea and problems with digestion of fats, but usually has no other major side effects. The gall bladder is situated immediately under the liver and attached to the small intestine by a series of ducts.

Because of the liver's vital roles in metabolic processes and blood filtration any failure of the liver is life-threatening. While it is unique among the organs for its ability to regenerate from a partial-organ transplant from a living donor, such transplants still require human donors. Unlike techniques such as heart-valve replacement where pig or cow heart-valves are used successfully, the liver is much more likely to suffer immune rejection in the host.

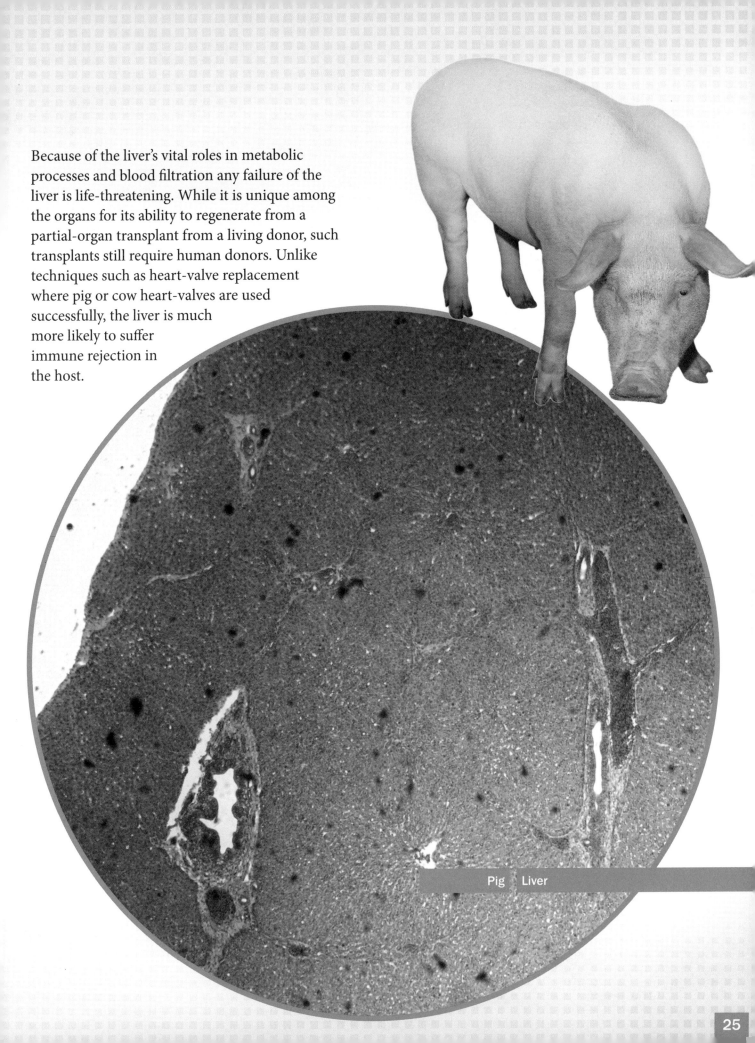

Pig | Liver

The central nervous system consists of the brain and spinal cord in vertebrates. Insects have simpler similar structures with a ventral nerve cord. The primary function of the spinal cord is to provide communication between the brain and peripheral sensory and motor neurons.

Rabbit   Spinal Cord

In the peripheral nervous system, a cluster of sensory nerve cell bodies is called a ganglion. The peripheral nerve fibers themselves, called axons, do not contain cell bodies and may be over 1 meter long.

Rabbit   Ganglion

Arteries and veins share similar basic structures: three layers known as the tunica intima, tunica media, and tunica adventitia. The tunica intima is the innermost layer: simple squamous epithelium attached to a thin membrane which provides a smooth surface that blood platelets won't stick to. The tunica media and adventitia make up the blood vessel wall and are made of smooth muscle, elastic fibers, and irregular connective tissues.

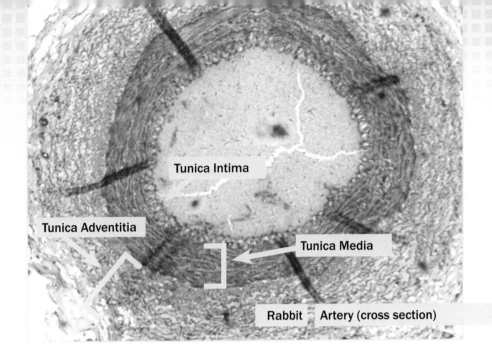

Tunica Intima

Tunica Adventitia

Tunica Media

Rabbit Artery (cross section)

Arteries lead away from the heart and carry blood under higher pressure than veins. They have thicker walls and tend to hold their shape when drained, unlike veins which collapse. The major vein in the body, the vena cava, is the final vein leading into the heart carrying deoxygenated blood to the heart for circulation back to the lungs before being circulated throughout the body. Many animals have several vena cava: one for the upper body, and one or two for the lower body.

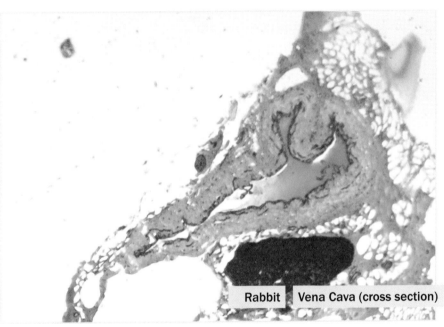

Rabbit Vena Cava (cross section)

While veins and arteries carry blood throughout the body, that wouldn't do much good if the blood didn't carry nutrients and oxygen for cells to use. Lungs are where all air-breathing animals exchange oxygen and carbon dioxide through tiny air pockets known as alveoli. The alveoli are surrounded by tiny blood vessels called capillaries that bring the blood close to the air being breathed so that gas exchange can occur through thin cells lining the alveoli called pneumocytes.

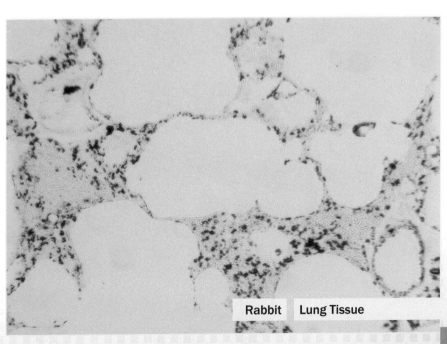

Rabbit Lung Tissue

Hookworms, pinworms and roundworms are all common names for species of nematodes: a broad classification that includes some 25,000 unique variations. Nematodes can be found in almost any environment, and are slender worms ranging in length from a tenth of a millimeter to five centimeters for **free-living** species. **Parasitic** forms may exceed 1 meter in length since they are protected by the host organism.

Common features in nematodes are an oral cavity, a simple digestive system running straight through the length of the worm and a hydrostatic skeletal structure. They also have a simple nervous system with a nerve ring near the head serving as a brain and four nerves running the length of the body for muscular control and sensing. Most nematode species are **dioecious**, having distinct genders and reproducing sexually.

While people are generally familiar with the **parasitic** organisms as pests to be eradicated from family pets or themselves, many forms of nematode are predators that attack agricultural pests and other insects. Undesirable **free-living** nematodes are managed by **antagonist** fungi and plants.

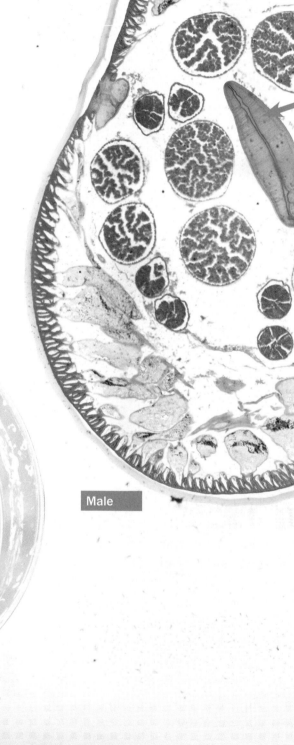

Male

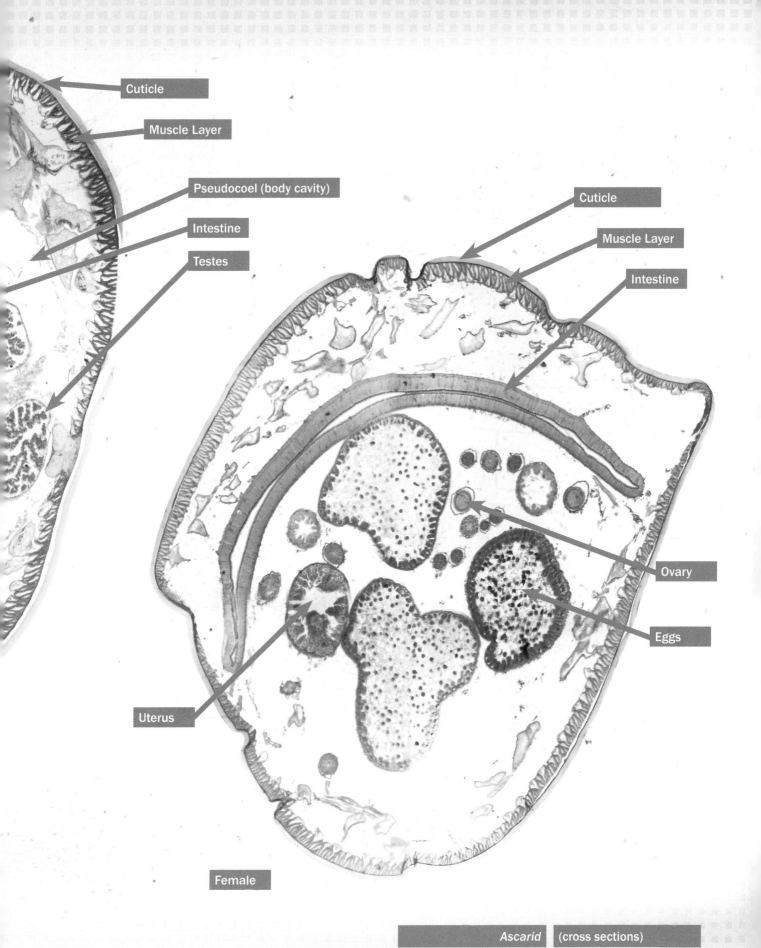

Cuticle

Muscle Layer

Pseudocoel (body cavity)

Intestine

Testes

Cuticle

Muscle Layer

Intestine

Ovary

Eggs

Uterus

Female

Ascarid (cross sections)

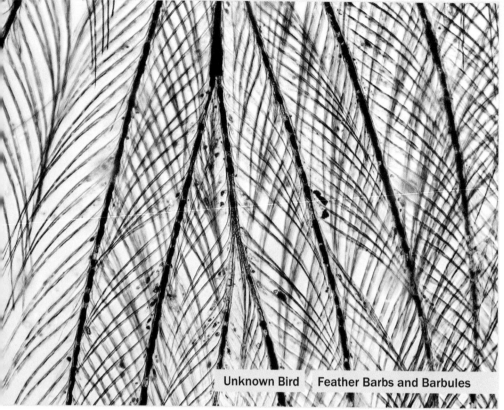

Unknown Bird | Feather Barbs and Barbules

Feathers have several major parts:

1. The rachis, or shaft, extending the length of the feather
2. Barbs, extending from each side of the rachis
3. Barbules, branching off of the barbs. In flight feathers, these have tiny hooks called barbicels connecting them together.
4. The hollow shaft, called a calamus, or quill.

Down feathers for insulation lack barbicels, allowing the barbules to appear fluffy and separate from each other.

In most birds, the feathers are waterproof, except for a few species like cormorants, whose feathers absorb water so they can swim submerged more easily.

Feather coloration has several purposes: camouflage and decoration among them. A dull-colored bird may be hiding from predators, or hiding itself from its prey. A brightly colored peacock, on the other hand, uses its brilliant plumage to attract a mate and in territorial displays. Some birds even have colors in the ultraviolet range that human beings can't see.

Peacock | Feather Barbules and Barbicels

Humans have used feathers for a variety of purposes over the years. Writing quills were crafted from the shafts of large feathers from around the 7th century through the early 19th century.

The original wooden golf balls were succeeded in the early 17th century by a feather-stuffed leather "featherie" that took several hours to make.

Down-lined jackets and sleeping bags are still in use today for their superior insulation and light weight when dry.

Peacock   Feather

Throughout the animal kingdom we see consistency of cell types and overall designs, but also unique variations such as the compounds that give blood its color and the differences in the barbules between smooth flight feathers and fluffy down.

All animals have a nervous system and the basic function and construction of their individual nerve cells is similar. In spite of this, the design of the nervous system as a whole is different among various animals, and uniquely suited to the size and structure of the individual species it is a part of: larger organisms have a central brain and spinal cord with nerves branching off to the extremities, while smaller creatures like nematodes have a simple nerve cluster or ring and several main nerves.

Similar designs are visible throughout other systems as well, like the increasing complexity of the circulatory system in organisms with higher blood flow requirements. The subdivision of the vena in certain species with multiple large veins merging the returning blood flow shows a well-designed system that is only as complex as it needs to be to support each organism.

Internal organs also showcase this design, and while some organs such as the gall bladder are not necessary for survival, all the organs work together to maintain the delicate balance required for life. Even in the absence of some of them the other systems strive to compensate, and although the organism is no longer functioning optimally, it is frequently able to continue living without radical changes to its behavior.

I praise you, for I am fearfully and wonderfully made. Wonderful are your works; my soul knows it very well.

Psalm 139:14

KINGDOM

SAMPLES

► INSECTS

Bee   Wing

Bee   Middle Leg

Bees have two pairs of wings that work together to provide lift, beating at speeds of 230 times per second. They stay in rhythm because a row of small hooks called *hamuli* along the leading edge of the rear wing hook onto the larger front wing and keep the wings moving at the same rate.

Bees fly up to 15 miles per hour and may travel up to 5 miles from their hive.

Bees have three pairs of flexible legs with six segments each. Every foot has claws for grasping and sticky pads for landing on slippery surfaces. They also have taste sensors on their feet like flies.

Worker bees have specialized rear legs with combs for collecting pollen off abdominal hair and carrying it back to the hive.

Bee   Rear Wing (featuring Hamuli)

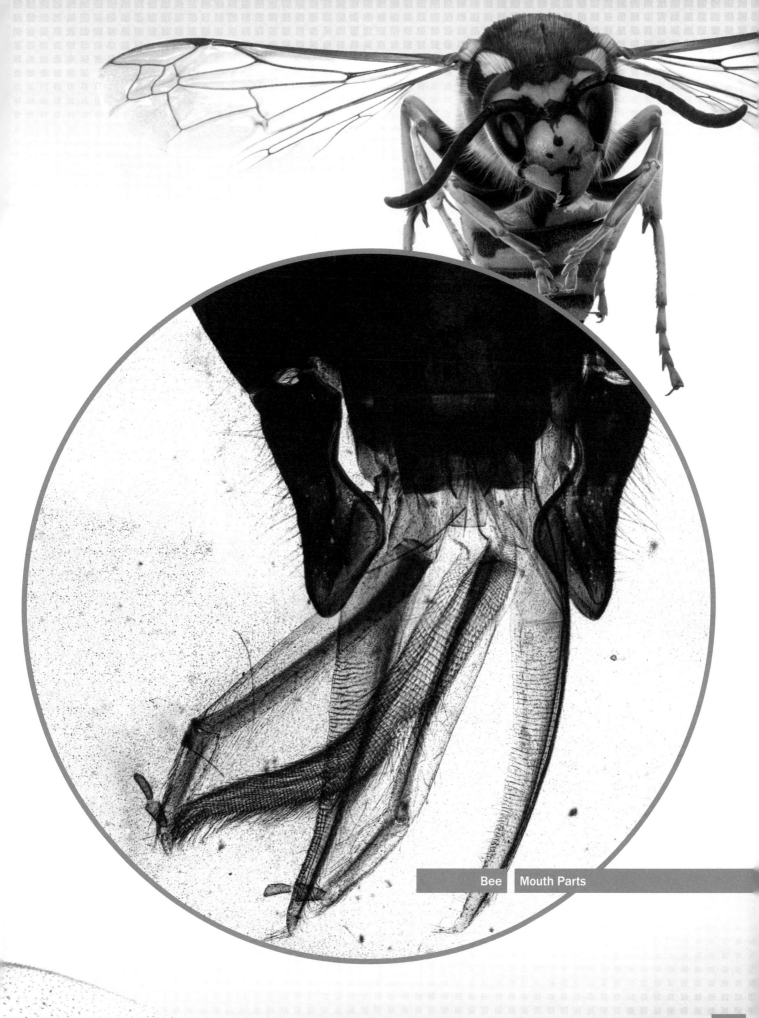

Bee | Mouth Parts

Butterfly legs, like bee legs, have six jointed segments and **chemoreceptors** on the lower portions for smelling and tasting. The segments are named the coxa, femur, trochanter, tibia, pretarsus, and tarsus. Butterflies have some form of claw on the tarsus as well. Butterfly antennae are clublike in appearance with segments growing larger as they grow further from the head. The antennae have chemoreceptors that the butterfly relies on for locating food and **pheromones** from other members of its species when mating.

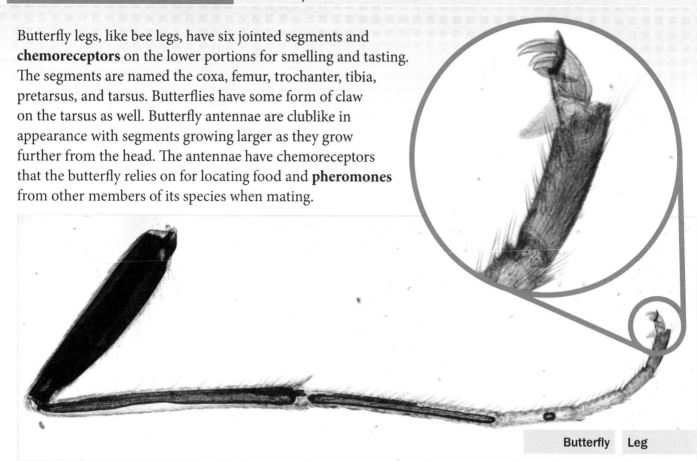

Butterfly   Leg

Butterfly   Antenna

The scientific name of a butterfly, *Lepidoptera*, literally means "scale winged". Shown below, a butterfly's wing is covered in hundreds of tiny scales, which contribute to a butterfly's brilliant coloration and improve its efficiency while flying. Scales also grow on the head, thorax and abdomen.

Butterflies' brilliant coloration comes from the scattering of light off the scales, not the color of the scales themselves, and the scales are loosely attached to the wing, pulling free without damage to the wing when the butterfly is trapped in a spider web. This increases the chances that a butterfly will survive without being permanently trapped. Body scales also trap air and insulate the body of the butterfly, allowing it to stay active in moderately cooler temperatures, although they still require plenty of basking to maintain an optimal 82 degree body temperature.

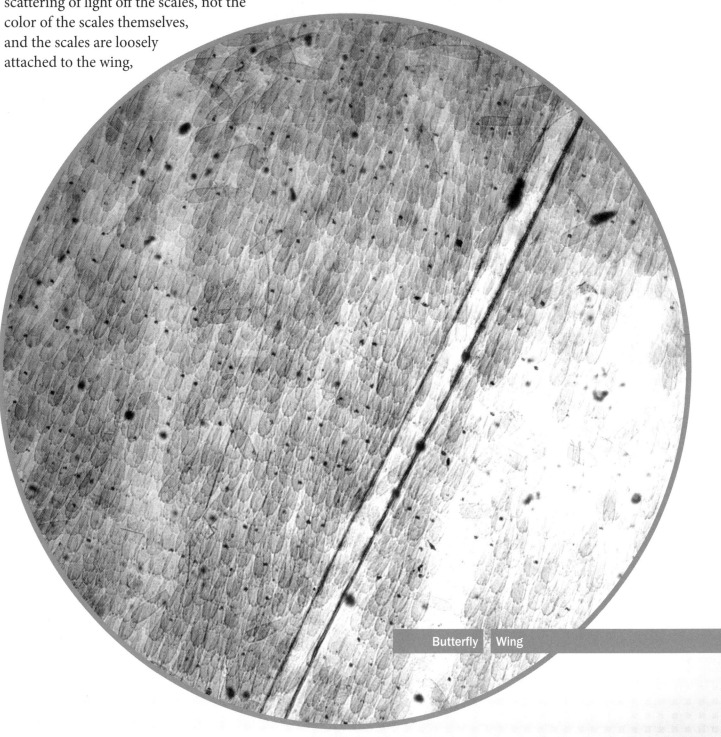

Butterfly | Wing

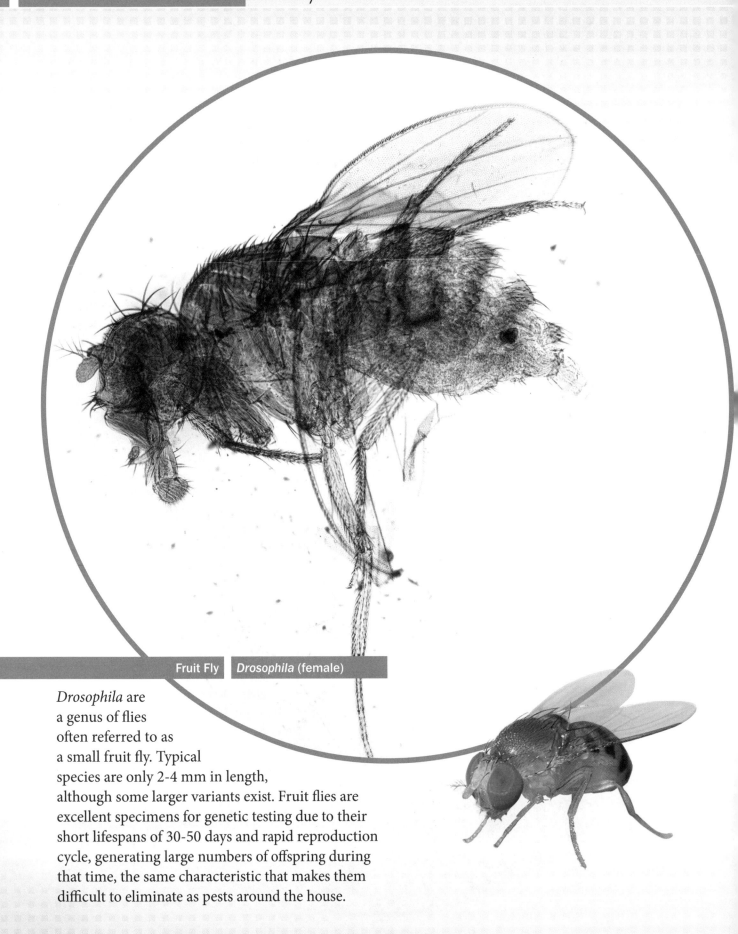

Fruit Fly | *Drosophila* (female)

Drosophila are
a genus of flies
often referred to as
a small fruit fly. Typical
species are only 2-4 mm in length,
although some larger variants exist. Fruit flies are
excellent specimens for genetic testing due to their
short lifespans of 30-50 days and rapid reproduction
cycle, generating large numbers of offspring during
that time, the same characteristic that makes them
difficult to eliminate as pests around the house.

Fruit flies lay 10-20 eggs at a time in rich food sources such as rotting fruit, although in leaves or other poorer **substrates** they may only lay a couple of eggs. These eggs hatch quickly, within a day or two, producing **larvae** (or maggots) that feed on yeasts and microorganisms in the decaying matter. Development of the larvae may take anywhere from one to eight weeks depending on environmental conditions.

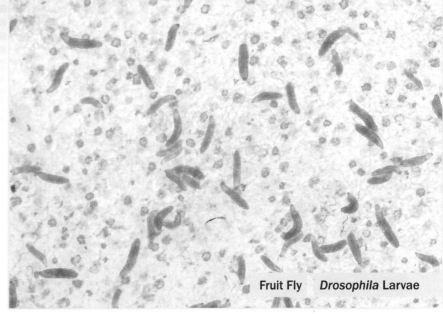

Fruit Fly   *Drosophila* Larvae

Fruit Fly   *Drosophila* Chrysalis

After maturing sufficiently as a larva, the young fruit fly forms a **chrysalis**, transforming into a **pupa** in order to make the transition from larva to adult fly. This period of protected development lasts approximately six days before a fully mature fly emerges from the chrysalis.

Mature fruit flies live as little as 10 days after emerging from their pupal form. Even a laboratory-kept sample under ideal conditions will only live a few months. Because of this, they mate rapidly within a day of emergence and reproduce quickly. **Model organism** *drosophila melanogaster*, studied because it is a good representative of the species, may lay up to 2,000 eggs during its brief adult life.

Fruit Fly   *Drosophila* (male)

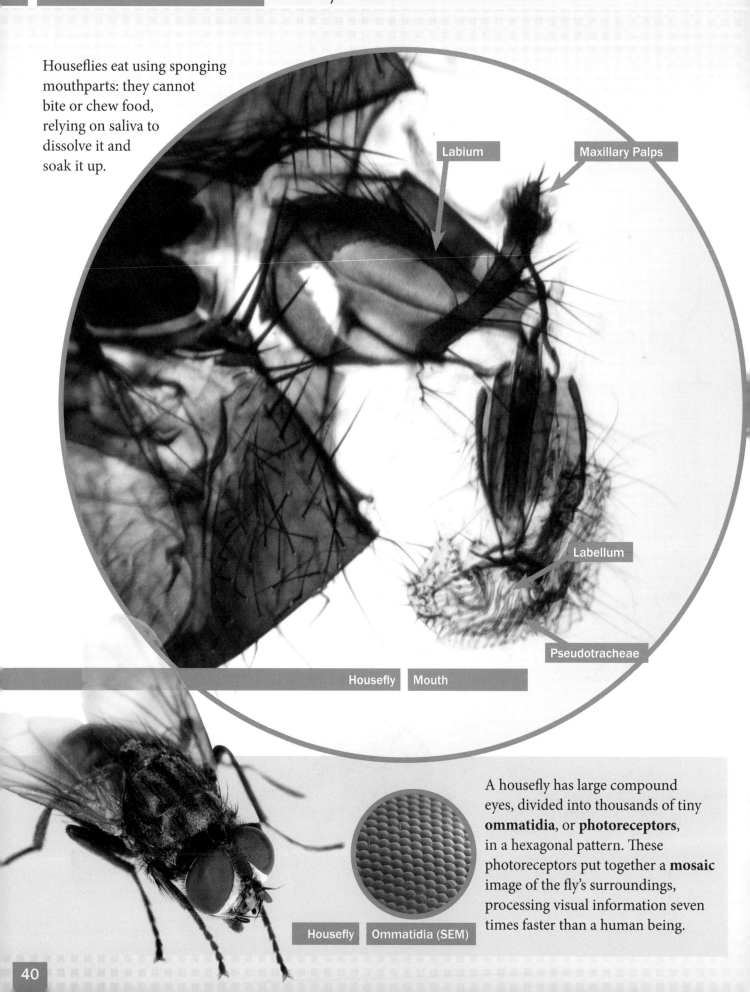

Houseflies eat using sponging mouthparts: they cannot bite or chew food, relying on saliva to dissolve it and soak it up.

Labium

Maxillary Palps

Labellum

Pseudotracheae

Housefly   Mouth

A housefly has large compound eyes, divided into thousands of tiny **ommatidia**, or **photoreceptors**, in a hexagonal pattern. These photoreceptors put together a **mosaic** image of the fly's surroundings, processing visual information seven times faster than a human being.

Housefly   Ommatidia (SEM)

Houseflies have a single pair of wings spanning roughly half an inch (13-15mm) and fly with a wingbeat of 200 strokes per minute. Combined with their relatively light body weight and quick reaction time, the housefly is extremely maneuverable, evading many predators in-flight with ease.

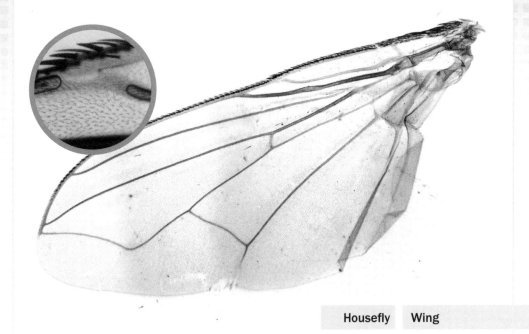

Housefly Wing

An adult housefly can lift about 10 mg (22 millionths of a pound), which is an impressive 50% of its body weight. Houseflies are six to seven mm long, with females being slightly larger than the males. They usually live two to three weeks, but may live up to two months.

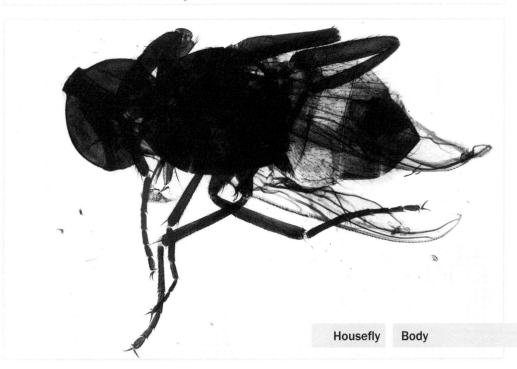

Housefly Body

Houseflies have a pair of claws at the end of each leg as well as two adhesive pads called pulvilli for walking up walls and on ceilings. Like bees and butterflies, houseflies also have **chemoreceptors** on their legs enabling a sense of taste, which is why they clean their legs so often.

Housefly Leg

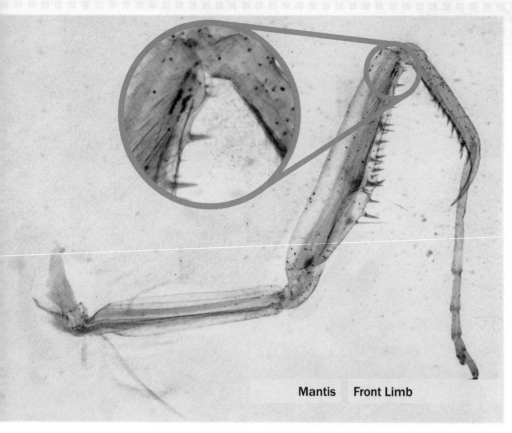

Mantis | Front Limb

The mantis family contains over 2,400 distinct species, but the most recognizable is probably the praying mantis. This distinctive creature keeps its forelimbs folded in a prayer-like posture before lashing out with extraordinary speed to snare its prey. The sharp spines on these forelimbs are ideally suited to this technique of active capture, and some mantises even pursue their prey, running them down instead of waiting for them to approach within range.

Mantises also use their **raptorial** forelimbs for defense, slashing at predators when cornered.

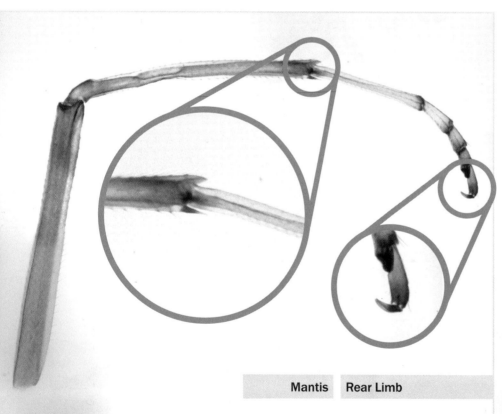

Mantis | Rear Limb

The rear limbs on a mantis are not as distinctive as the forelimbs, but are perfectly suited for locomotion with claws at the tips for grasping sticks and twigs. Mantises are capable of balancing upright on their rearmost pair of legs, using their middle legs as well as their "arms" in defense and territorial posturing.

Mantises have two pairs of wings: thin, delicate rear wings and leathery forewings which close over them for protection. Mantises generally fly at night, avoiding most birds that might prey on them.

Mantis Wing

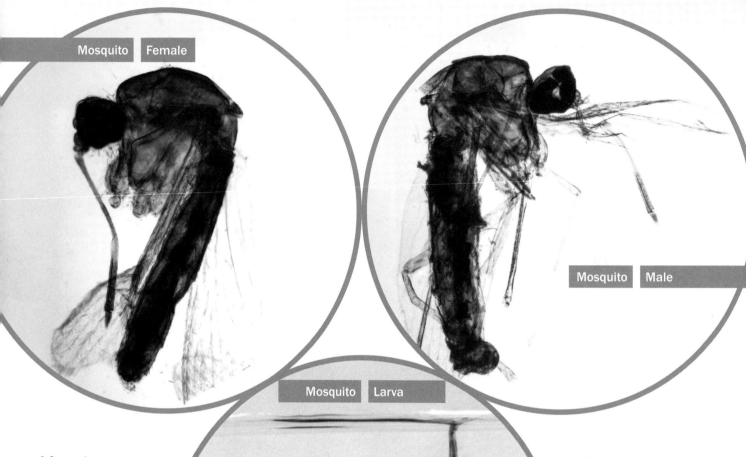

Mosquito | Female

Mosquito | Male

Mosquito | Larva

Mosquitoes are incredibly common and widespread with over 3,500 distinct species found worldwide. Most species have a similar life cycle: eggs are laid in the water, hatch into **larvae**, progress through several stages of **molts**, then metamorphose into **pupae** before emerging as adult mosquitoes.

The female mosquito, as the egg layer, is the primary nuisance to humankind and animals. While male mosquitoes subsist almost entirely on nectar and other sources of sugar during their week-long lifespan, the female mosquito requires dense volumes of iron and protein for egg production. Blood is the preferred source, often from insects and birds, but larger mammals become targets as the mosquito population rises and competition for food becomes a factor. This change in food sources creates a significant **disease vector** as organisms ingested earlier by the mosquito are now transmitted during feeding to other host species.

The female mosquito's mouthparts are uniquely suited to obtaining blood. The elongated **proboscis** consists of a protective sheath, a set of **mandibles** and **maxillae** used to saw their way through the host's skin rather than piercing through it, a **hypopharynx** for injecting **anticoagulant**-containing saliva, and a hollow **labrum** that draws the blood up.

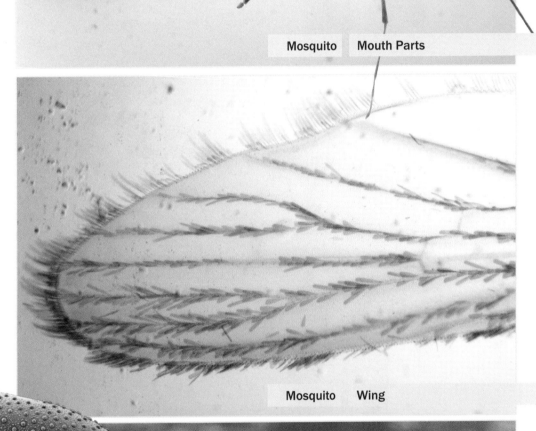

Mosquito | Mouth Parts

Mosquitoes have a single pair of wings and regularly beat their wings between 450 and 600 times per second, making a high-pitched whine in flight. The mosquito flies by moving its wings in a unique figure-eight pattern, taking advantage of **air vortices** from the previous stroke to increase efficiency.

Mosquito | Wing

Mosquito | Egg (SEM)

Mosquitoes lay their eggs in moist areas: most either singly or in "rafts" of eggs in standing water, but some species lay them in mud. Eggs generally do not hatch until they have been submerged in water, but can usually survive dry spells before hatching by entering a form of dormancy called **diapause**.

Mosquito | Egg Raft

Silkworm | Larva (side view)

Domesticated silkworms are the larva of *Bombyx mori*, the domestic silkmoth. Throughout its long life of selective breeding, this species has been carefully groomed to produce high-quality silk, and in the process has lost the ability for sustained flight as well as most natural pigmentation. The domestic silkmoth is no longer capable of survival in the wild.

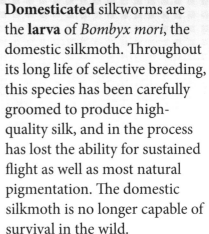

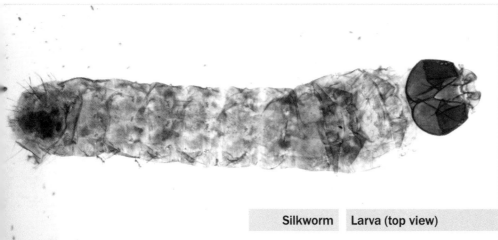

Silkworm | Larva (top view)

Instead, domestic silkmoths are kept in captivity and lay about 500 eggs during their lifetime, which hatch in a couple of weeks. The larvae eat continuously, preferring white mulberry leaves. They will continue to eat through four **moltings** until cocooning themselves, at which point they will **pupate**, forming the valuable cocoon they were raised for.

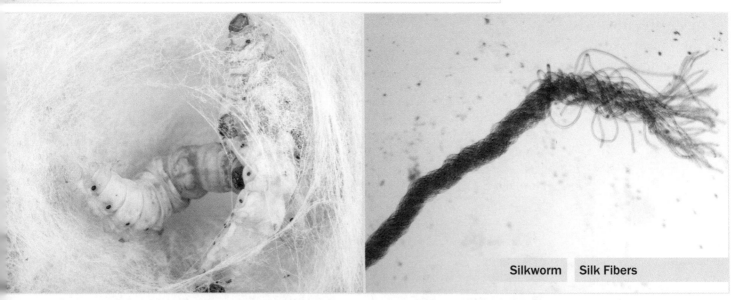

Silkworm | Silk Fibers

A silkworm's cocoon is made of a single thread of raw silk 10 μm (0.0004 in) in diameter and up to a mile long! The cocoon is boiled to kill the larva inside, preventing it from destroying the silk thread by emerging. This also dissolves the sericin gum cementing the filaments together. 3,000 to 5,000 cocoons go into each pound of finished silk: enough to make a single Japanese kimono. This large-scale farming of silkworms is called sericulture and has been documented in China since before Confucius (about 350 B.C.). The process was a closely guarded secret for centuries, but has slowly spread to other parts of the world over the last two thousand years.

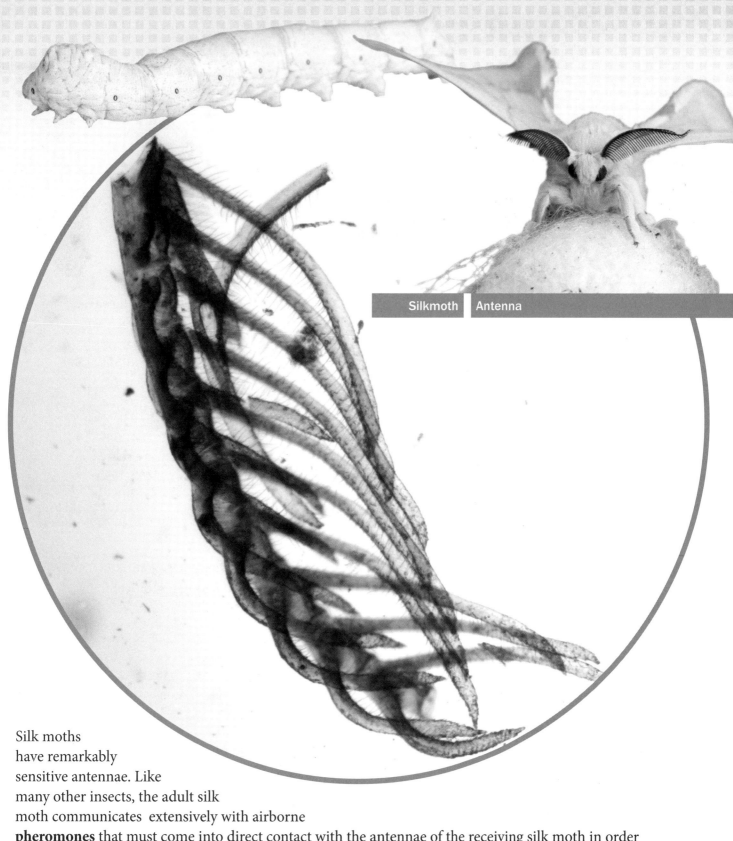

Silkmoth | Antenna

Silk moths
have remarkably
sensitive antennae. Like
many other insects, the adult silk
moth communicates extensively with airborne
pheromones that must come into direct contact with the antennae of the receiving silk moth in order
to be detected. The comblike structure of a silk moth's antennae vastly increases the amount of air that
passes through and is sampled for pheromones compared to other species' stalk-like designs.

Silk moths in the wild have very low population densities, so this ability to communicate and attract a
mate over long distances is crucial to their survival. In fact, it is estimated that 70 percent of the sensilla,
or sensory **chemoreceptors**, on a giant male silk moth's antennae are exclusively dedicated to detecting
females. This makes sense because adult silk moths have no working mouths. This final stage in their life
cycle is dedicated to finding a mate and reproducing before they starve in a couple of weeks.

As we explored the insect kingdom, we saw many similarities in overall structure. Most insects grow through a four-phase lifecyle: beginning from an egg and passing through **larval** and **pupal** phases before emerging as a mature adult. Some species such as dragonflies have a single phase called a nymph, characterized by the lack of developed wings, instead of the larval and pupal stages. These insects go through multiple successive **molts** as the nymph grows into its adult form.

Most insects have large compound eyes that excel at detecting motion, and some also have simple, single lensed eyes (ocelli) that are designed for a specific purpose like sun orientation or detecting motion out of their primary field of vision. Other sensory organs such as antennae are shared among most insects as well, but with widely varying forms. As we saw on the silk moth, their antennae have a unique comblike structure that is exquisitely designed to enhance sensitivity to airborne signals. Butterflies, bees, and other insects have smell and taste receptors on their antennae, and frequently on their legs as well.

Even the mouthparts of insects have amazing variety. From flies with their sponging mouthparts suited to sucking up liquified food, butterflies with long, flexible strawlike tongues for feeding from flowers, and mosquitoes with an arsenal of tools for cutting through tough skins, whether animal or plant, we see exhibits of excellent design and unique capabilities.

> Then God said, "Let the earth bring forth living creatures after their kind: cattle and creeping things and beasts of the earth after their kind"; and it was so. God made the beasts of the earth after their kind, and the cattle after their kind, and everything that creeps on the ground after its kind; and God saw that it was good.
>
> **Genesis 1:24–25**

KINGDOM | SAMPLES

➡ PLANTS & FUNGI

Plant cells are very similar to animal cells, in that they contain **cytoplasm**, a **nucleus**, and **mitochondrions** for energy production. Clearly, plants and animals aren't exactly the same, and there are some definite differences between plant and animal cells as well.

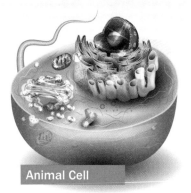
Animal Cell

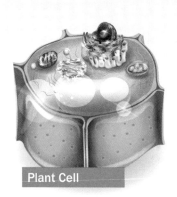
Plant Cell

Plant cells have a **cellulose cell wall** instead of a **cell membrane**, which gives each distinct cell a definite shape. Unlike animal cells which will burst in water or a dilute solution, plant cell walls resist the entry of excess water and prevent the cell from bursting. Cells also have a large fluid-filled area called a **vacuole**. When a cell's vacuole is full of water the cell is firm and provides strong support for the plant. If the cells dry out they lose their rigidity and the plant wilts.

There are three major physical structures in plants: leaves, stems, and roots. The leaves tend to be large and thin, with a high ratio of surface area to volume. This is important for photosynthesis, which requires a large amount of energy absorbed from sunlight. The high surface area also improves the efficiency of gas diffusion through the stomata, which are small mouthlike openings on the underside of leaves that regulate gas exchange.

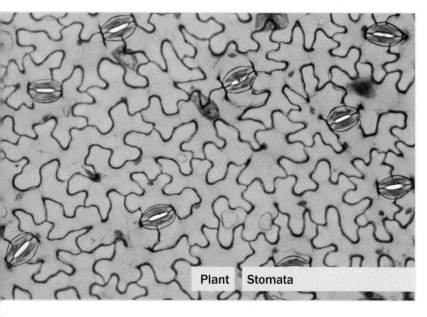
Plant   Stomata

Cells in leaves where photosynthesis takes place also contain **chlorophyll** in tiny **organelles** called **chloroplasts**. These photosynthesizing cells use energy from sunlight and carbon dioxide from the atmosphere to form sugars and O_2, which are then used for **respiration** in the plant. Excess O_2 from this process is diffused into the atmosphere.

The second major physical structure in plants is the stem, made of long cylindrical fibers. Veins run throughout the stems, comprised of **xylem**, a woody vascular tissue that carries water and dissolved nutrients upwards from the roots, and **phloem**, which moves sugars downward from the leaves. This circulation allows the cells in the roots to **respire**, in spite of not being able to conduct photosynthesis to generate food.

Plant   Stem (longitudinal section)

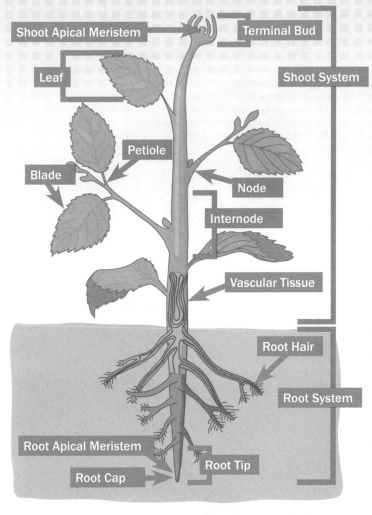

- Shoot Apical Meristem
- Terminal Bud
- Leaf
- Shoot System
- Petiole
- Blade
- Node
- Internode
- Vascular Tissue
- Root Hair
- Root System
- Root Apical Meristem
- Root Tip
- Root Cap

The last major plant structure is the roots. These branch extensively through the soil, and are classified according to one of two major types. Plants with a taproot look like the image at the bottom left, with a single central root and several rootlets branching off of it. Plants with fibrous or adventitious roots look more like grass when it is pulled up: a network of fine fibers spreading out close to the surface. These plants help prevent soil erosion, but are limited to drawing water and minerals from close to the surface.

Both root systems serve to anchor plants in the soil and have tiny root hairs with relatively large surface areas for absorption. Roots grow by dividing the cells at the tip of the root, known as the root cap, with the cells immediately behind the root cap elongating into a mature root. Some rarer forms of roots grow above ground. These are known as aerial and prop roots and are seen in sandy and marshy areas where **indigenous** plants require more support. Other unique root forms provide nutrient stores for the plant, and in turn for humans gathering them, such as beets, carrots, and sweet potatoes.

Fibrous Root

Tap Root

Prop Root

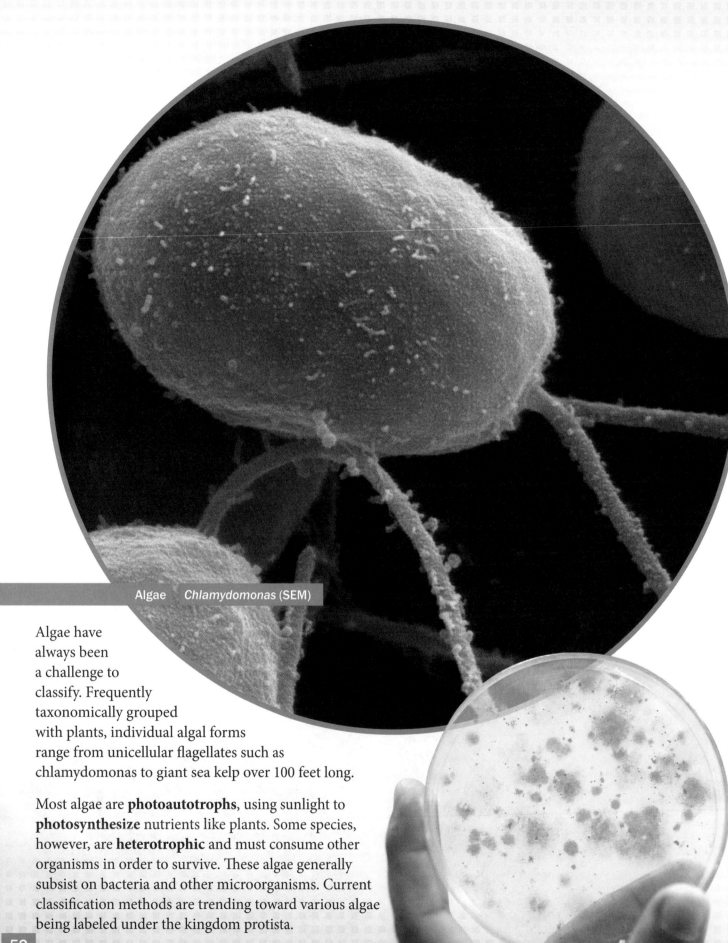

Algae   *Chlamydomonas* (SEM)

Algae have
always been
a challenge to
classify. Frequently
taxonomically grouped
with plants, individual algal forms
range from unicellular flagellates such as
chlamydomonas to giant sea kelp over 100 feet long.

Most algae are **photoautotrophs**, using sunlight to
photosynthesize nutrients like plants. Some species,
however, are **heterotrophic** and must consume other
organisms in order to survive. These algae generally
subsist on bacteria and other microorganisms. Current
classification methods are trending toward various algae
being labeled under the kingdom protista.

Multicellular algae grow in various forms ranging from the cellularly uniform to the **differentiated**. Structures like the long, spiraling chains of *Spirogyra* and spherical colonies like *Volvox* are formed by uniform cells grouping together. Seaweeds and kelps, on the other hand, have differentiated cells in various parts of the structure that have distinct appearances and functions. One example is their rootlike holdfasts that hold the organism in place, but unlike roots do not absorb nutrients.

Photosynthetic algal forms vary in color primarily depending on which pigment they contain. **Chlorophyll** and phycoerythrin, contained in green and red algae respectively, both perform the same function of absorbing light energy to fuel photosynthesis, but work efficiently with different wavelengths of light. These different pigments allow some marine algae to grow deeper than 800 feet, where only reddish algae can absorb the bluish green light reaching those depths.

Algae are used in a variety of consumer and industrial products. Many species of seaweed and kelp, such as *Porphyra* and *Palmaria palmata* are edible. Derivatives of the **cell walls** known as alginates, agars, and carrageenans are used as thickeners, fillings, and stabilizers in both foods and cosmetics. Some algae classified as diatoms have cell walls made of opaline silica, which becomes diatomite when the algae die and the cells drift together. The fine pores make diatomite an excellent filter.

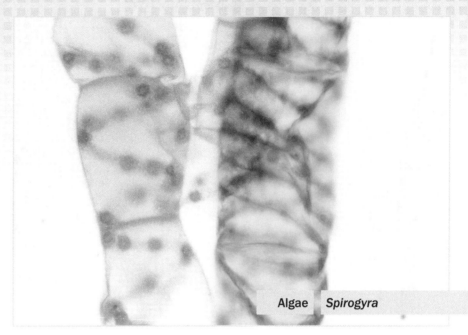

Algae   *Spirogyra*

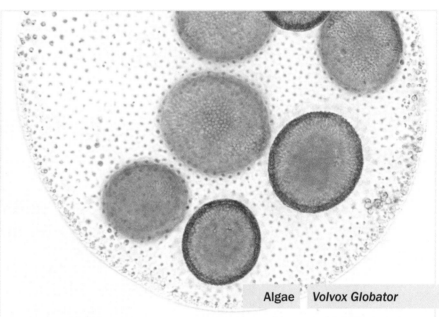

Algae   *Volvox Globator*

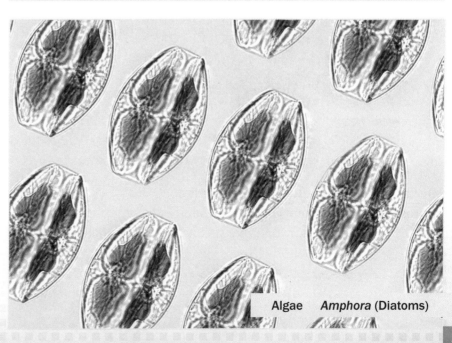

Algae   *Amphora* (Diatoms)

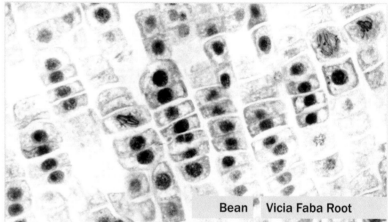

Bean | Vicia Faba Root

The broad bean, also known as a fava or faba bean, is one of many members of the legume family, which includes a wide variety of food crops like lentils, peas, and coffee. The term "bean" refers to common plants that produce large edible seeds or seedpods. Most beans contain nutrients that make them valuable sources of both dietary protein as well as vitamins and minerals.

Bean plants are valuable crops as well because they form a **symbiotic** relationship with bacteria like *Rhizobium* that allows them to be nitrogen fixing, converting nitrogen from the air into compounds that plants can use. This prepares cultivated fields for other plants that must absorb most of their nitrogen directly from the soil like cabbage, kale, and cauliflower.

Legumes sprout from a single seed, forming a small shoot and a taproot. During the early stages of growth, root hairs form and rhizobia bacteria living in the soil attach themselves to the root hairs, prompting the formation of round nodules on the roots where the bacteria live for the life of the plant.

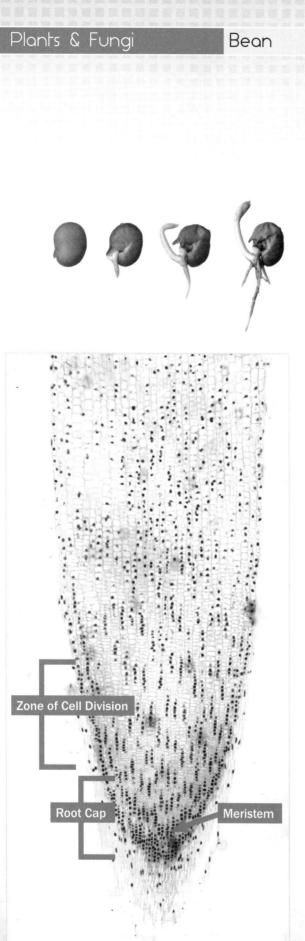

Zone of Cell Division

Root Cap

Meristem

Bean | *Vicia faba* Root (Tip)

The root structure is divided into three major sections: the epidermis, cortex, and stele. The epidermis is the tough, outermost layer of cells, protecting the root from abrasion. The cortex provides most of the mass of the root, and the stele contains the **xylem** and **phloem**, which function as the plant's circulatory system.

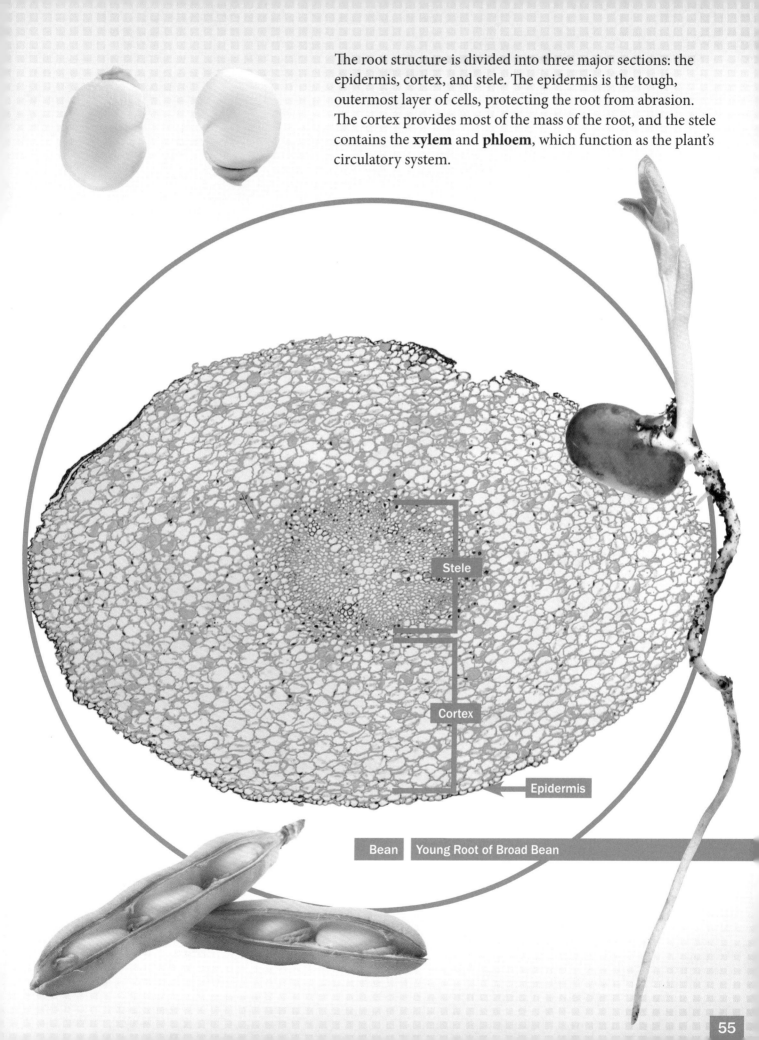

Stele

Cortex

Epidermis

Bean | Young Root of Broad Bean

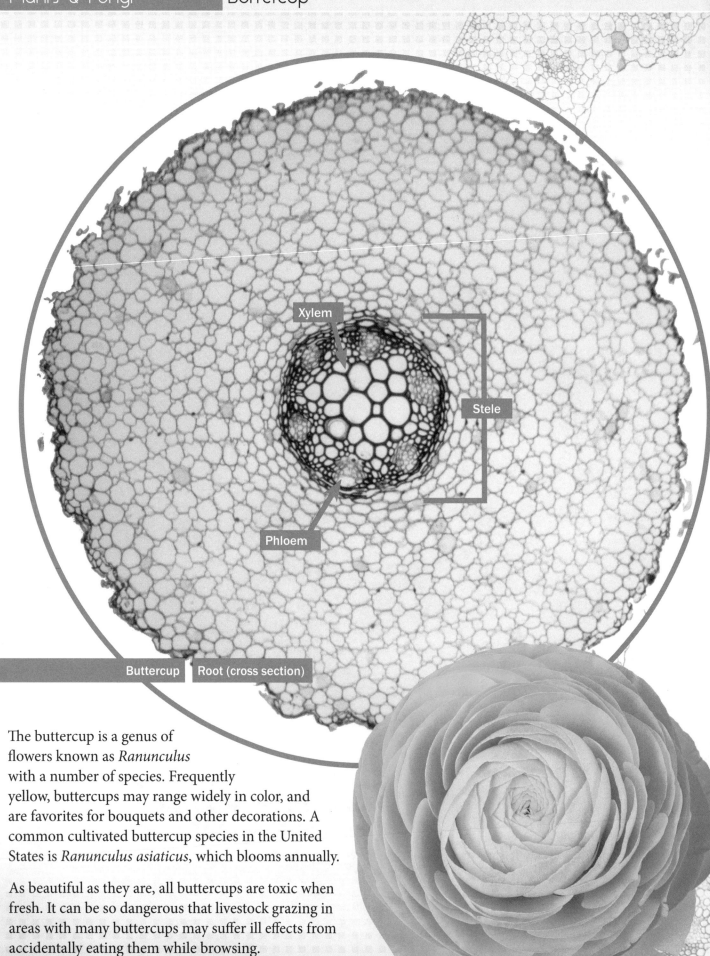

Xylem

Stele

Phloem

Buttercup   Root (cross section)

The buttercup is a genus of flowers known as *Ranunculus* with a number of species. Frequently yellow, buttercups may range widely in color, and are favorites for bouquets and other decorations. A common cultivated buttercup species in the United States is *Ranunculus asiaticus*, which blooms annually.

As beautiful as they are, all buttercups are toxic when fresh. It can be so dangerous that livestock grazing in areas with many buttercups may suffer ill effects from accidentally eating them while browsing.

The buttercup is a dicot, having two **cotyledons** (or embryonic leaves) in the seed. A way to identify it as a dicot when it is grown is to take a cross section of the root and see that the **xylem** creates a central area in the root with a ring of **phloem** tissue bundles around the perimeter of the stele. Dicots also tend to have webs of veins in their leaves and flower petals in multiples of four or five, while monocots have parallel veins in their leaves and petals in multiples of three.

Stems of dicots and monocots have similar structures to their roots: the vascular tissue bundles of dicots are still arranged around the outside of the stem. The major difference is that the xylem tissue is now separated into bundles along with the phloem instead of being a central mass with bundles of phloem around it. The phloem is still closer to the periphery of the stem with the xylem arranged to the interior. These bundles of vascular tissue transport liquids and nutrients, as well as providing support through fibrous **sclerenchyma** cells with **lignin**-thickened **cell walls**.

The other two tissues in plant stems are the epidermis and ground tissues. The epidermis is the thin outer layer protecting the plant from damage and excessive moisture loss. Ground tissue comprises the filler in the cortex and pith, composed primarily of **parenchyma** metabolic cells for **photosynthesis**, but also containing some sclerenchyma and **collenchyma** cells. Unlike sclerenchyma cells, which die at maturity and form woody tissue, collenchyma cells form living connective tissue and are usually found close to the epidermis.

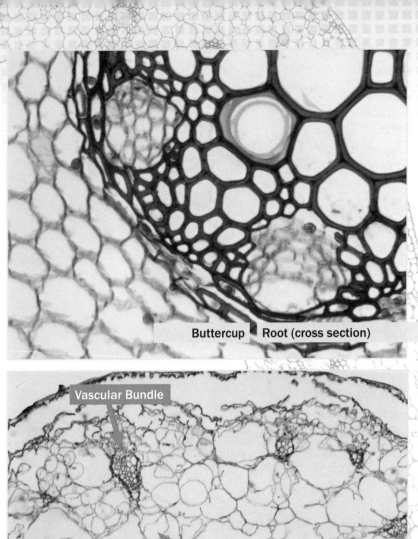

Buttercup   Root (cross section)

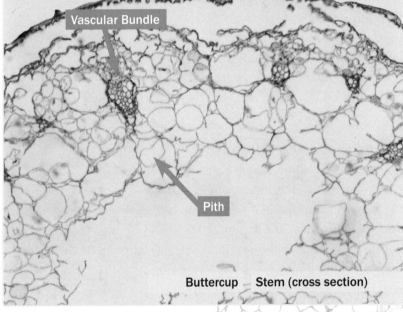

Vascular Bundle

Pith

Buttercup   Stem (cross section)

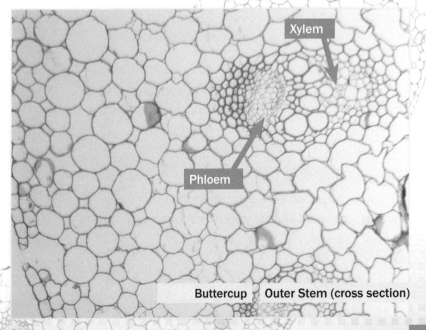

Xylem

Phloem

Buttercup   Outer Stem (cross section)

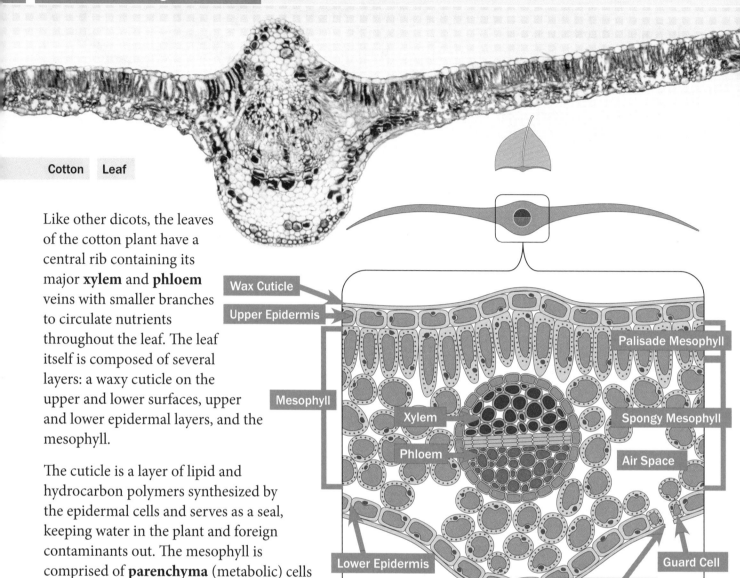

Cotton | Leaf

Xylem

Phloem

Wax Cuticle

Upper Epidermis

Palisade Mesophyll

Mesophyll

Spongy Mesophyll

Air Space

Lower Epidermis

Guard Cell

Wax Cuticle

Stoma

Like other dicots, the leaves of the cotton plant have a central rib containing its major **xylem** and **phloem** veins with smaller branches to circulate nutrients throughout the leaf. The leaf itself is composed of several layers: a waxy cuticle on the upper and lower surfaces, upper and lower epidermal layers, and the mesophyll.

The cuticle is a layer of lipid and hydrocarbon polymers synthesized by the epidermal cells and serves as a seal, keeping water in the plant and foreign contaminants out. The mesophyll is comprised of **parenchyma** (metabolic) cells and subdivided into two sublayers. Palisade parenchyma forms a dense layer in the upper surface of the leaf with tightly packed cells for efficient absorption of sunlight. Spongy parenchyma comprises a porous layer in the lower portion.

The lower epidermal layer has a number of stomata, the orifices through which gas exchange is regulated. These open into substomatal chambers in the leaf: gaps in the spongy parenchyma that function very similarly to alveoli in lungs, exposing a large cellular surface area for gas exchange.

While the leaves of the cotton plant are typical, the reproduction cycle is what sets it apart. During its 160 day growing period, a cotton plant will grow small white or yellow flowers. These will pollinate, turn pinkish, and then wither away, exposing a cotton boll underneath. This boll contains an average of 32 cotton seeds padded in a bundle of approximately 250,000 fibers. These fibers, ranging from 1/8 inch to 2 inches in length, are gathered and processed into various textiles, with the longer fibers being more valuable for their strength and quality of finished product.

Cotton fibers are oval in shape and have a convoluted, or twisted appearance, with about 150 revolutions per inch. These twists increase friction among the fibers and enable the fabrication of fine, strong threads.

Cotton | Raw Cotton Fibers

Processing raw cotton involves drying the fibers, ginning the cotton (separating the fibers from the seeds) carding the fibers into parallel strands, and spinning them into thread. A process of chemically treating cotton textiles, called mercerization, is commonly used to pre-shrink the treated fabric and increase its resistance to tearing. The process changes the shape of the cotton fibers to a rounder profile, which makes the resulting cloth less absorbent, but also less prone to staining and damage.

Cotton | Stem

Most animals reproduce sexually, with some organisms like frogs, starfish, and hydras being capable of **asexual reproduction** as needed. Plants, however, have multiple forms throughout their life cycle. This is termed alternation of generations and described in terms of **haploid** and **diploid** cycles with the **somatic** cells having one and two sets of **chromosomes** respectively durinåg these phases. Ferns are a good example of this, having free living **gametophytes** that are easily studied.

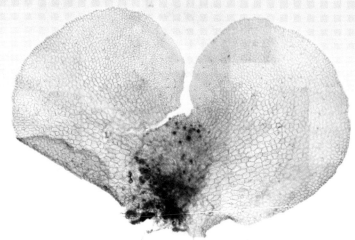

Fern   Gametophyte (Prothallus)

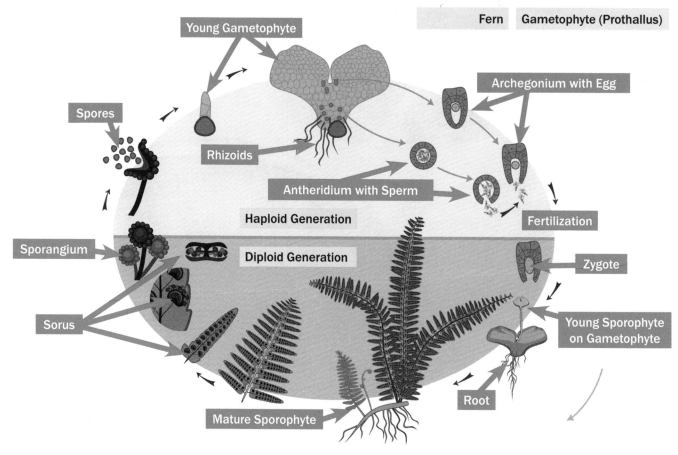

During the haploid phase of a fern's generational cycle, a spore will grow into a single-cell thick organism called a gametophyte, or prothallus. This organism has elongated cells known as **rhizoids** that anchor it to the soil. Most gametophytes contain both **antheridium** (male) and **archegonium** (female) sex organs, which mature at different times to increase the chances of cross-fertilization, although some ferns produce two separate genders of spores, and therefore discrete genders of gametophytes.

After fertilization of the archegonium, a **zygote** cell is formed. This begins the diploid cycle of growth as a **sporophyte** emerges from the gametophyte with the **zygote** dividing through **mitosis**, eventually growing into a recognizable mature fern. This fern will grow clusters of **sporangia** called sori on the underside of its leaves, in which each sporangium will undergo **meiosis**, releasing numerous spores with half the chromosomes of the parent plant, starting the haploid cycle again.

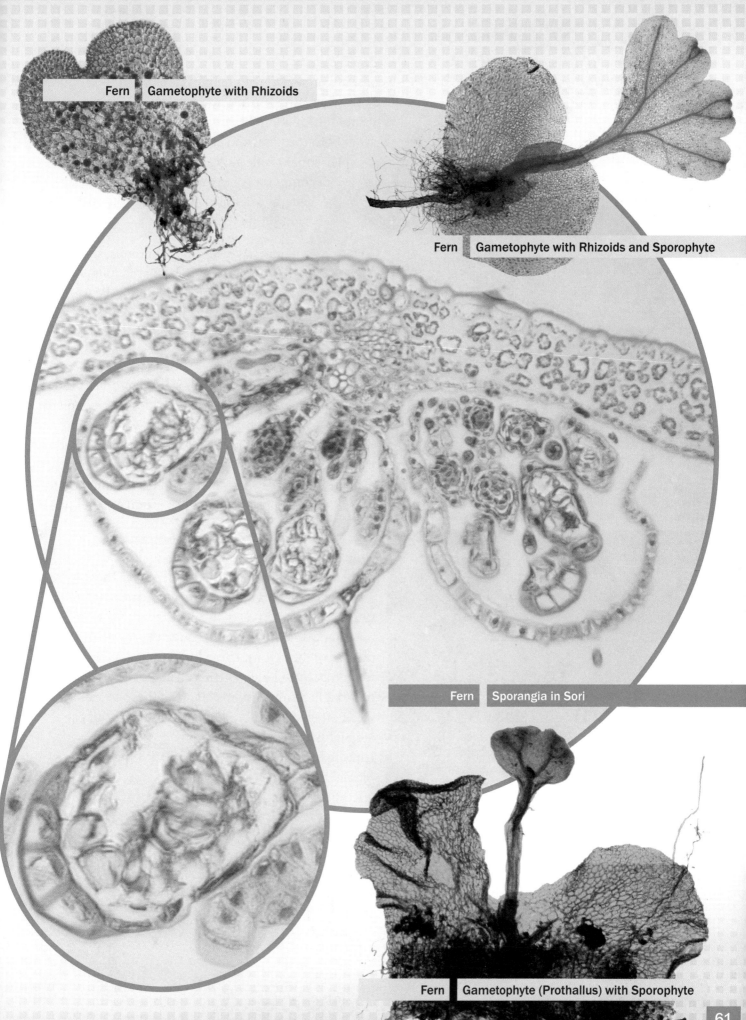

Fern | Gametophyte with Rhizoids

Fern | Gametophyte with Rhizoids and Sporophyte

Fern | Sporangia in Sori

Fern | Gametophyte (Prothallus) with Sporophyte

Fungi are a common organism with many different forms. Originally classified as a plant, after further research fungi were classified with a unique kingdom because they display characteristics from both the plant and animal domains. Fungi play a major role along with bacteria as natural decomposers of organic material, since they generally dissolve molecules for food and do not **photosynthesize**.

Fungi have three major groups: single-celled microscopic, multicellular filamentous, and macroscopic filamentous. Few of these forms are **motile**, most spreading through growth in colonies instead of actively moving around.

The single-celled microscopic fungi, yeasts, have been commonly used for millennia for the byproducts of their consumption of sugars: carbon dioxide and alcohols. The carbon dioxide makes dough rise when baking, and yeasts left to grow over a long period of time produce wines, beers, and other fermented beverages. More recently, the use of yeasts as a cost-effective method to process wood chips and waste paper into ethanol fuel is being explored.

Unlike yeasts, all other forms of fungus grow multicellular filaments known as **hyphae**. These form a network known as **mycelium** which is considered a single organism. Ends of the hyphae strands form spores (conidia), which then spread out and grow to increase the size of the colony. Large (macroscopic) forms of fungi that form large **fruiting** structures are commonly referred to as mushrooms. Some of these are edible, while others like the *Amanita phalloides* (death cap) are highly toxic. Non-fruiting multicellular filamentous forms are called molds.

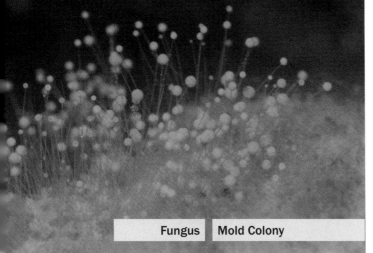

| Fungus | Yeast Budding (SEM) |

| Fungus | Mold Colony |

| Fungus | *Aspergillus* |

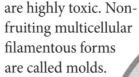

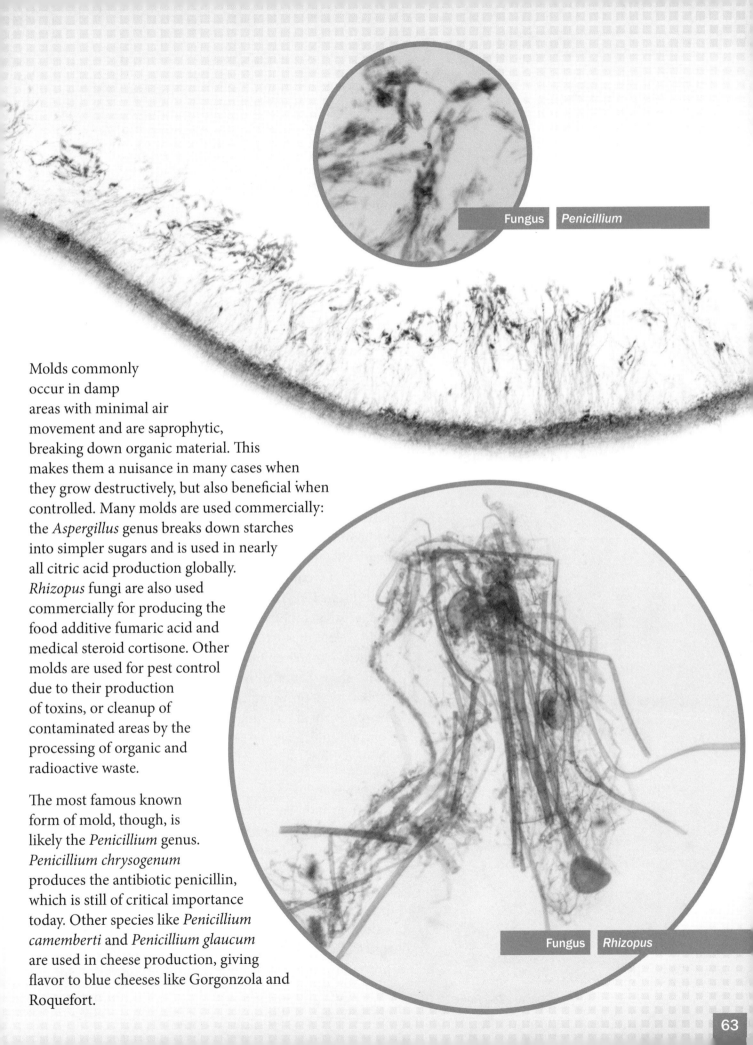

Fungus   *Penicillium*

Molds commonly occur in damp areas with minimal air movement and are saprophytic, breaking down organic material. This makes them a nuisance in many cases when they grow destructively, but also beneficial when controlled. Many molds are used commercially: the *Aspergillus* genus breaks down starches into simpler sugars and is used in nearly all citric acid production globally. *Rhizopus* fungi are also used commercially for producing the food additive fumaric acid and medical steroid cortisone. Other molds are used for pest control due to their production of toxins, or cleanup of contaminated areas by the processing of organic and radioactive waste.

The most famous known form of mold, though, is likely the *Penicillium* genus. *Penicillium chrysogenum* produces the antibiotic penicillin, which is still of critical importance today. Other species like *Penicillium camemberti* and *Penicillium glaucum* are used in cheese production, giving flavor to blue cheeses like Gorgonzola and Roquefort.

Fungus   *Rhizopus*

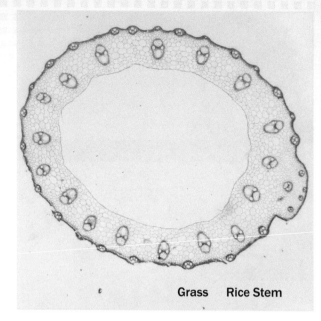

Grass   Rice Stem

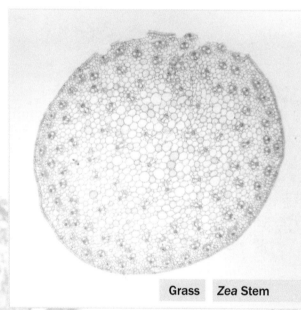

Grass   Zea Stem

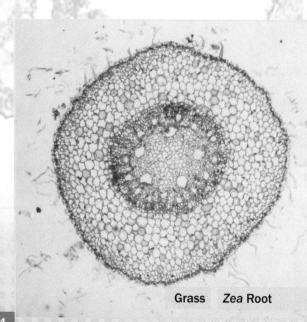

Grass   Zea Root

Grass   Zea Root

Grass is found all around the earth in various forms. It isn't just represented by the varieties found in lawns though: there are many diverse species of grass like rice, wheat, corn, sugarcane and bamboo. Grasses are monocots as seen by their vascular tissue patterns. The stems usually have bundles of **xylem** and **phloem** together distributed throughout the stem and roots with individual bundles distributed in a ring. There are a few exceptions though: wheat and rice stems look more like dicots with a ring of vascular bundles around the periphery, but they have leaves with parallel veins and seeds with a single **cotyledon.**

Grasses have a wide variety of uses. Bamboo has comparable strength characteristics to both concrete (**compressive**) and steel (**tensile**). These characteristics combined with its light weight make it an ideal construction material in areas where it grows abundantly. Not only that, but it grows quickly: some species of giant bamboo can grow as much as 3 feet in a 24-hour period.

Turf grasses are used in landscaping and play an important role in preventing soil erosion with their branching, fibrous root systems. These types of grasses spread rapidly because they can not only reproduce by spreading seeds, but also by **rhizomes** producing nodes that sprout another plant. In addition to preventing soil erosion, root masses from dead grasses decompose to enrich the soil.

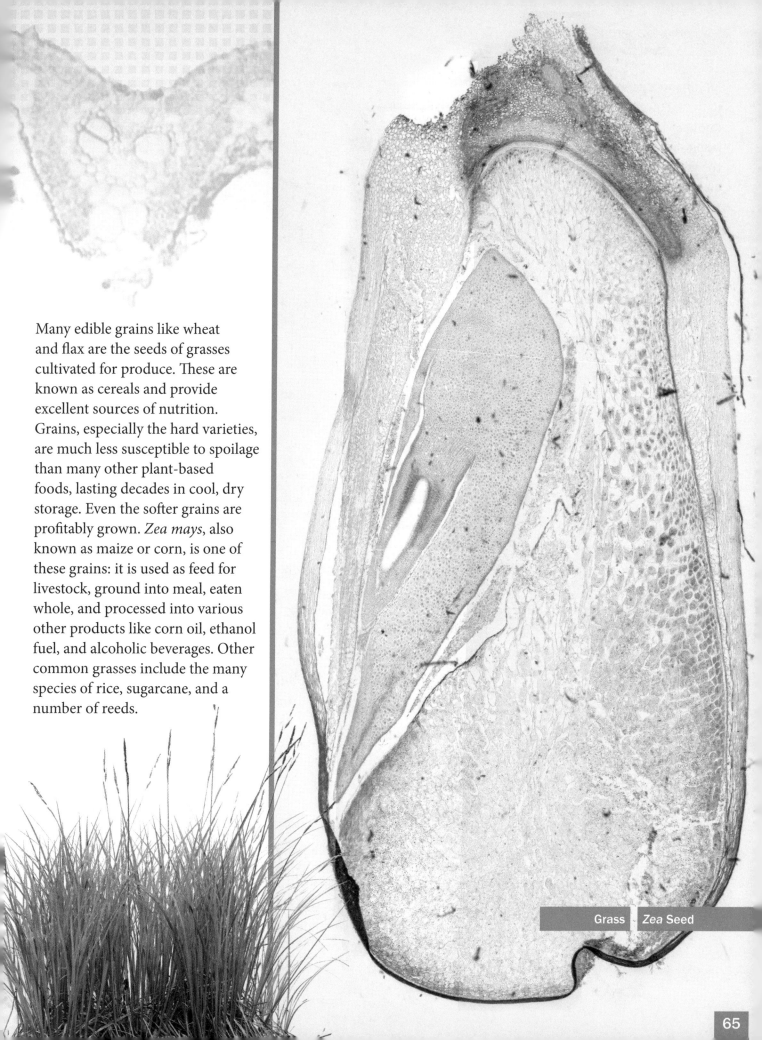

Many edible grains like wheat and flax are the seeds of grasses cultivated for produce. These are known as cereals and provide excellent sources of nutrition. Grains, especially the hard varieties, are much less susceptible to spoilage than many other plant-based foods, lasting decades in cool, dry storage. Even the softer grains are profitably grown. *Zea mays*, also known as maize or corn, is one of these grains: it is used as feed for livestock, ground into meal, eaten whole, and processed into various other products like corn oil, ethanol fuel, and alcoholic beverages. Other common grasses include the many species of rice, sugarcane, and a number of reeds.

Grass | Zea Seed

Lilium are a genus of flowering **perennial** monocots native to the northern hemisphere. Commonly known as lilies, these plants are prized for their beauty and variety in gardens. Members of the genus are capable of both sexual and **asexual** reproduction, and frequently reproduce asexually in cultivation because of their long maturation time of two to six years before flowering.

During asexual reproduction, lilium naturally extend **rhizomes** from the parent plant's bulb. These shoot off and form new bulbs that grow into a separate plant. Cultivated leaf cuttings and scraped **bulb scales** can also grow into copies of their parent.

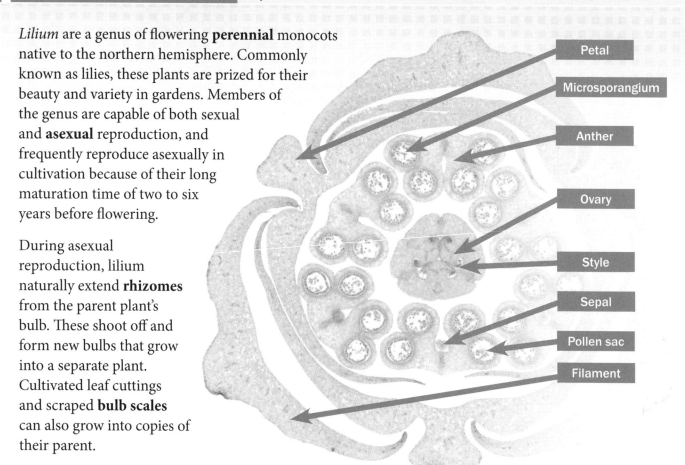

Petal

Microsporangium

Anther

Ovary

Style

Sepal

Pollen sac

Filament

Sexual reproduction in lilium can occur even within a single plant since, like most plants, they are **hermaphroditic**. The male reproductive structures are known as **androecium**, or **stamen**, containing a stalklike filament, a pod at the tip known as the **anther**, and a number of **microsporangium** in the **anther** where **meiosis** occurs to produce pollen grains. The female counterpart to the androecium is the **gynoecium**, or **pistil**, which consists of a **stigma** at the top where pollen lands, a **style** through which it grows downward, and an ovary with **ovules** which will form new seeds when fertilized.

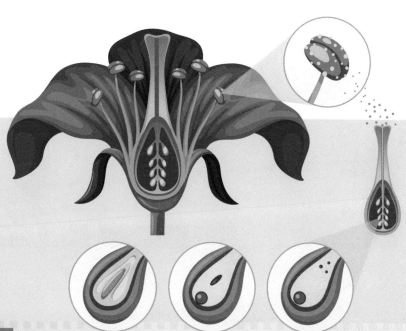

Pollination occurs when the pollen produced at the stamen of a flower reaches the ovary in the pistil of a flower with a sperm cell. Both the ovule and the sperm are produced through meiosis and carry half of the genetic material required for a new plant.

This image illustrates the stamen (orange structure) producing pollen, which then is transported to a stigma at the center of a flower. There it grows a pollen tube down through the style to an ovary where it transfers a sperm cell into an ovule, forming a new seed that will grow into a plant after **germination**.

Lilium are entomophilous, or insect-loving plants. Plants classified this way tend to have large, vibrant flowers to attract pollinating insects and produce relatively large, heavy grains of pollen. Typically non-flowering plants that produce small, light pollen grains are described as anemophilous, or wind-loving, because a breeze can spread their pollen. In both cases, the **anther** opens and releases the pollen grains when the **microsporangia** (pollen sacs) are mature, as seen below.

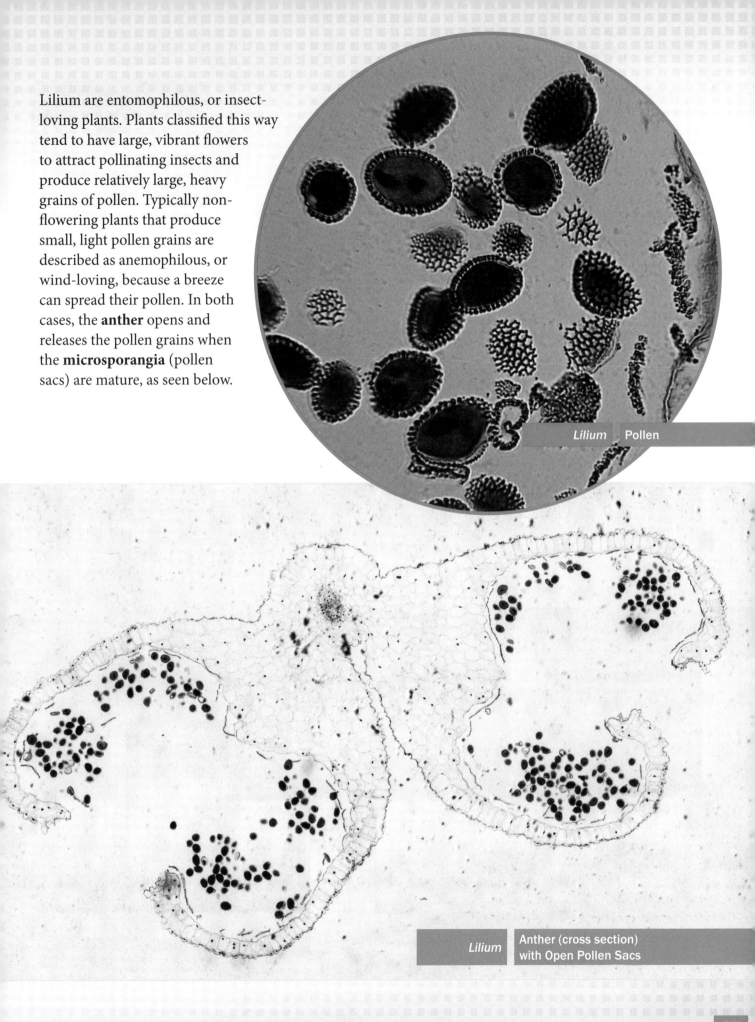

Lilium Pollen

Lilium Anther (cross section) with Open Pollen Sacs

Marchantia polymorpha, or the common liverwort, is a member of the Marchantia genus of plants which are classified as bryophytes along with mosses and hornworts. Marchantia grow differently than many common plants with **differentiated** structures, instead growing thalli, which are a mass of undifferentiated tissue. While there may be some minor differences in appearance, the cells of a thallus are all of the same unspecialized type, with no separation into distinct organs like veins of **xylem** and **phloem**.

Marchantia | Thallus with Gemma Cups

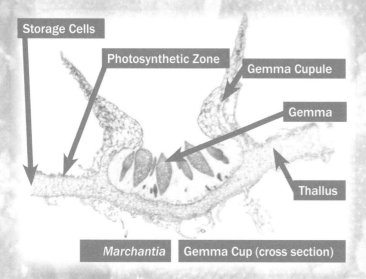

Storage Cells

Photosynthetic Zone

Gemma Cupule

Gemma

Thallus

Lateral Notch

Marchantia | Gemma Cup (cross section)

Marchantia | Gemma

As **dioecious** plants, Marchantia reproduce both **asexually** and sexually with distinct genders. Asexual reproduction occurs when the **gametophyte's** thallus forms gemmae. These small cellular bodies or buds cluster in formations called gemma cups and are separated from the parent plant by rainfall. When a mature gemmae reaches the ground, it sends out a number of **rhizoids** and then forms two thalli in opposing directions from the lateral notches to grow into a complete new **gametophyte**.

For sexual reproduction, the male plant forms specialized stalks, or **antheridiophores**, where the sperm-producing **antheridia** grow. Female plants have similar structures for the production of **ovum** called **archegoniophores** and **archegonia**. Upon fertilization, the gametophyte form's ovum becomes a **zygote** and forms a new **sporophyte** that begins producing spores while still attached to its parent. These spores disperse and develop into new gametophyte forms of the liverwort.

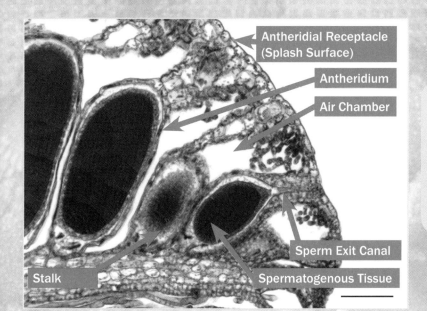

Antheridial Receptacle (Splash Surface)

Antheridium

Air Chamber

Sperm Exit Canal

Stalk

Spermatogenous Tissue

Marchantia antheridiophore (longitudinal section)

Underneath the top surface (splash platform) can be seen air chambers as well as antheridia.

Scale = 0.2mm.

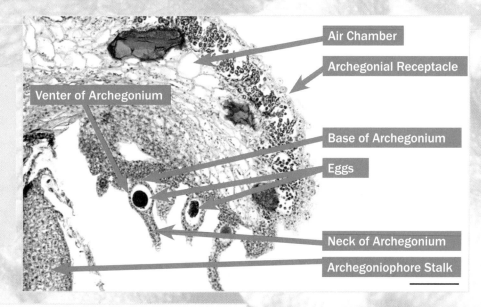

Air Chamber

Archegonial Receptacle

Venter of Archegonium

Base of Archegonium

Eggs

Neck of Archegonium

Archegoniophore Stalk

Marchantia archegonial receptacle (longitudinal section)

On the bottom surface can be seen the eggs within the venters, attached by the base of the archegonium with the necks pointing downwards.

Scale = 0.1mm.

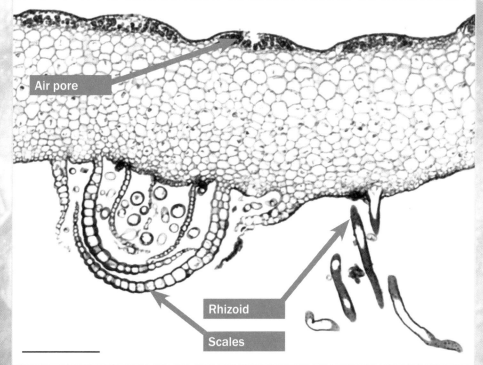

Air pore

Rhizoid

Scales

Marchantia thallus (transverse section)

On the upper surface, an air pore is visible.

The underside shows the protective scales and a rhizoid.

Scale = 0.2mm.

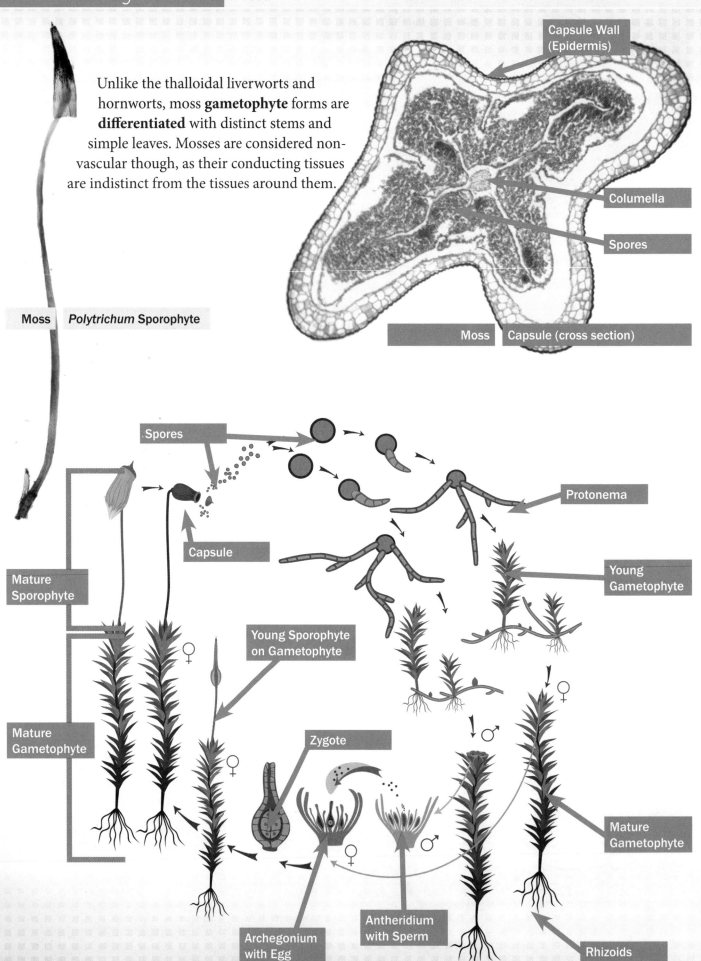

Unlike the thalloidal liverworts and hornworts, moss **gametophyte** forms are **differentiated** with distinct stems and simple leaves. Mosses are considered non-vascular though, as their conducting tissues are indistinct from the tissues around them.

Capsule Wall (Epidermis)

Columella

Spores

Moss | *Polytrichum* Sporophyte

Moss | Capsule (cross section)

Spores

Protonema

Capsule

Young Gametophyte

Mature Sporophyte

Young Sporophyte on Gametophyte

Mature Gametophyte

Zygote

Mature Gametophyte

Archegonium with Egg

Antheridium with Sperm

Rhizoids

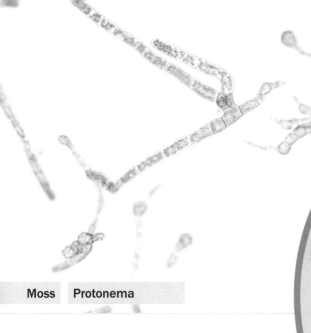

Moss Protonema

Mosses are small plants generally growing only an inch or two in height that spend most of their lives in the gametophyte form. The gametophyte phase begins when a moss spore germinates to produce a protonema. These will grow on most surfaces, and eventually transition to the gametophore that is typically recognized as the moss plant.

Moss *Polytrichum* Archegonia

Gametophytes may be either **dioecious** or **monoecious**, and produce **antheridia** and **archegonia** structures in order to reproduce sexually. Sperm cells of mosses are biflagellate, having two tendrils that they use to swim when immersed in water. A fertilized **ovum** in an archegonia will grow into a **sporophyte** with a capsule in the tip where spores mature and disperse three to six months later, returning to the beginning of the cycle. Some mosses can also reproduce **asexually** in their gametophyte form like liverworts by producing gemma.

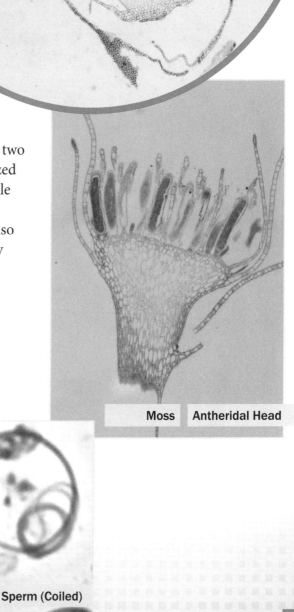

Moss Antheridal Head

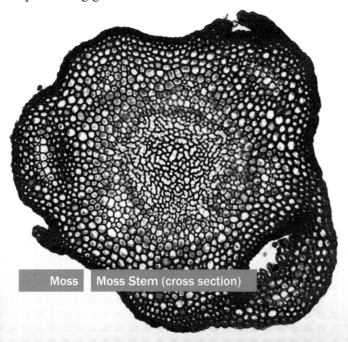

Moss Moss Stem (cross section)

Moss Sperm (Coiled)

Pine

Pine trees are vascular plants with long life cycles, typically living from 100-1000 years, with some species spanning millennia. Due to their vascular structure and long lifespans with stems forming trunks through secondary lateral growth, they provide a history of general climate conditions throughout their life. The science of dendrochronology is the practice of identifying and dating the age of trees from the patterns formed in their trunks by varying growth rates through the seasons.

These growth rings form when the rate of growth in the vascular **cambium** changes: slower growth rates during colder months or droughts form denser wood which shows up as a dark ring, while faster growth forms less dense, lighter areas. This means that it is possible for there to be multiple rings per year if there are phases of drought, but generally the rings follow annual cycles.

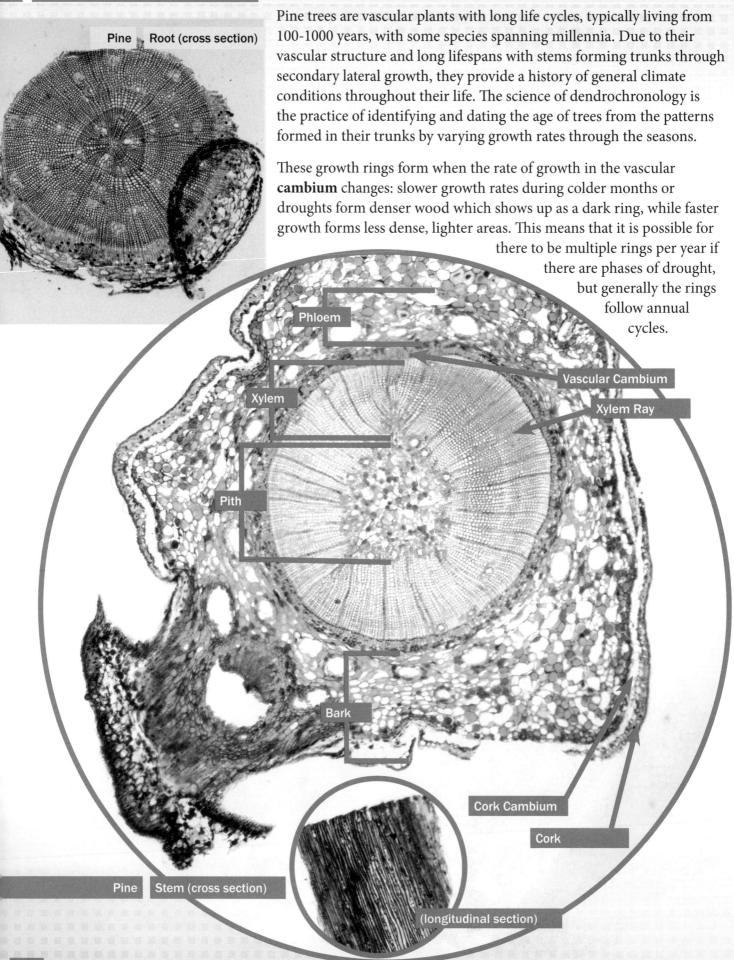

Pine | Root (cross section)

Phloem

Vascular Cambium

Xylem

Xylem Ray

Pith

Bark

Cork Cambium

Cork

Pine | Stem (cross section)

(longitudinal section)

As vascular plants, pines have well-developed circulatory systems, and their leaves show this as well. While small, pine needles share the same structures as larger flat leaves, including **xylem** and **phloem** vascular bundles, various **differentiated** tissue types and stomata for gas exchange.

Along with their needle-like leaves, pine trees reproduce by forming seed-bearing cones instead of fruit or flowers. These cones identify them as gymnosperms, or "naked seeds." Gymnosperms are **monoecious**, with the same plant growing both male and female reproductive organs. In most cases, the female cones are high up where wind-blown pollen can easily fertilize the **ovules**, while the male cones grow lower down.

Pine | Leaf (cross section)

Sunken Stoma

Mesophyll

Xylem

Phloem

Transfusion Tissue

Endodermis

Hypodermis

Epidermis

Resin Canals

In female (ovulate) cones, each scale contains two ovules, each of which contains a megaspore cell that will divide and eventually form an **archegonia**. A male (staminate) cone produces microspores, which in turn divide into pollen grains. These pollen grains are usually dispersed by the wind, and will land on the **scales** of an ovulate cone and grow pollen tubes to reach the egg enclosed in the archegonium. When fertilized, the egg will form an embryo and seed coat; this entire process may take several years as the growth of the pollen tube takes up to a year prior to fertilization taking place.

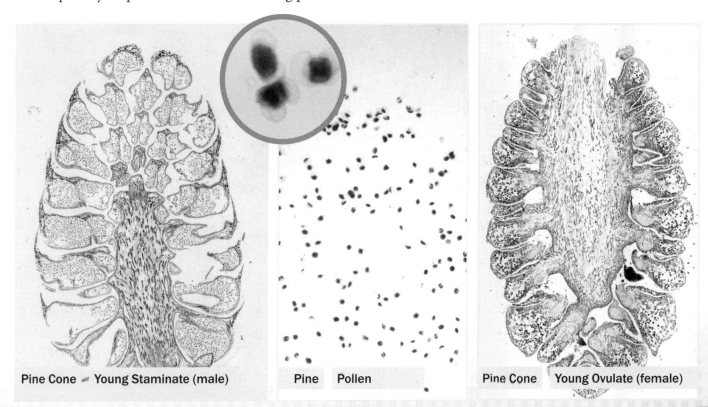

Pine Cone — Young Staminate (male)

Pine | Pollen

Pine Cone | Young Ovulate (female)

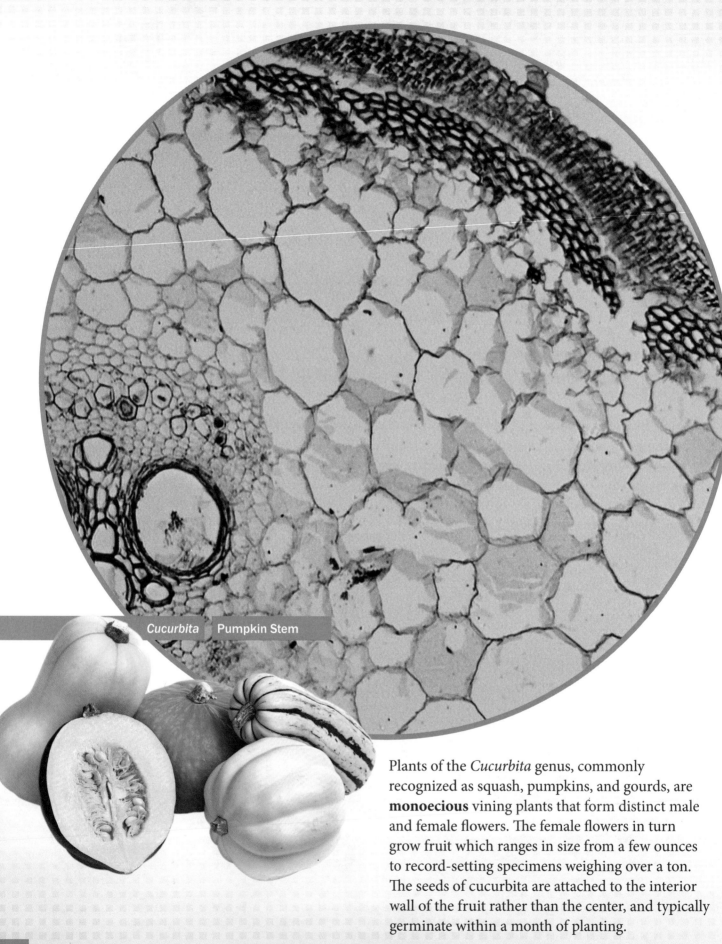

Cucurbita   Pumpkin Stem

Plants of the *Cucurbita* genus, commonly recognized as squash, pumpkins, and gourds, are **monoecious** vining plants that form distinct male and female flowers. The female flowers in turn grow fruit which ranges in size from a few ounces to record-setting specimens weighing over a ton. The seeds of cucurbita are attached to the interior wall of the fruit rather than the center, and typically germinate within a month of planting.

Cucurbita   Pollen

Pollination occurs chiefly via bees visiting cucurbita flowers, with the amount of pollen transferred directly influencing the number of seeds and thereby the size of the mature fruit as well.

Like other woody dicots, the vines of cucurbita have a structure similar to trees, where the ring of vascular bundles joins together around the periphery of the stem forming a continuous ring. This structure forms a number of layers from the epidermis at the outside inward to the pith.

The outer epidermis is composed of compact cells and may have stomata in a young stem. The next layer is the cortex which is divided into three zones: patches of **collenchyma** cells known as the hypodermis, a layer of **parenchyma** cells, and the starch sheath composed of larger cells. The stele, or central cylinder, contains the vascular bundles, supporting **sclerenchyma**, and some parenchyma tissue. The pith tends to disintegrate quickly in cucurbita, leaving a gap in the center of the stele.

The vascular bundles in cucurbita tend to be bicollateral: there is a ring of **xylem** with **cambium** and **phloem** on both sides. The vascular tissues contain a number of elongated cells, known as sieve tubes, which form conduits for water and nutrients to pass through.

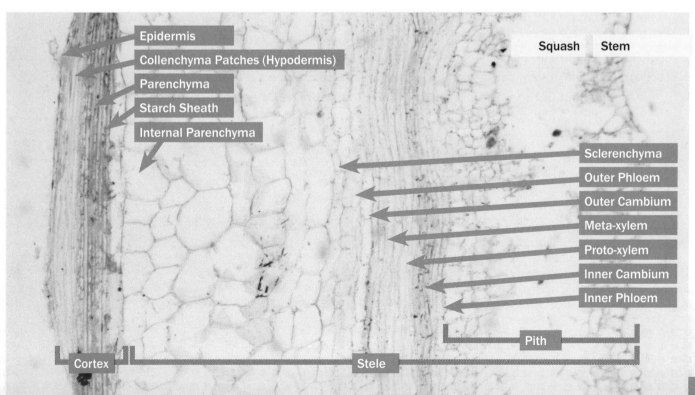

Epidermis
Collenchyma Patches (Hypodermis)
Parenchyma
Starch Sheath
Internal Parenchyma

Squash   Stem

Sclerenchyma
Outer Phloem
Outer Cambium
Meta-xylem
Proto-xylem
Inner Cambium
Inner Phloem

Pith

Cortex   Stele

Through our exploration of plants and fungi, we have seen a wide variety of sizes and appearances. From photosynthetic algae and moss to buttercups and pine trees, **chloroplasts** and the corresponding ability to generate food from sunlight and organic compounds are consistently found. This process establishes a global **symbiotic** relationship between plants and animals, as the carbon dioxide released during animal **respiration** becomes a primary food source for plants as they use the carbon to build sugars for food and release the oxygen.

The design of plant structures is amazing as well; even though they don't actively relocate themselves, plants still circulate fluid and food throughout their leaves and roots, quickly allocating resources among these areas. They will even reposition themselves to expose their leaves to sunlight more efficiently, and some store excess nutrients for periods of deprivation in their roots. These roots also show unique characteristics of God's design, with various forms like taproots and fibrous roots being matched well with the physical requirements of the plant for support and maintenance of its environment. Plant leaves also reflect this suitability for their environment, with leaves in drier areas having thicker waxy cuticles and less surface area to lose moisture from.

Even the complicated cycle of reproduction that plants go through is amazing evidence of design. The pollen of plants that produce flowers to attract insects is heavier, while other plants like pines produce their male cones lower down the tree so that they are less likely to self-fertilize when the wind disperses their pollen. Because many plants can also reproduce **asexually**, this means that they are less likely to die out due to isolation if conditions are unfavorable for sexual reproduction.

> I will put in the wilderness the cedar, the acacia, the myrtle, and the olive. I will set in the desert the cypress, the plane and the pine together, that they may see and know, may consider and understand together, that the hand of the Lord has done this, the Holy One of Israel has created it.
>
> **Isaiah 41:19–20**

KINGDOM
SAMPLES

➡ MICROORGANISMS

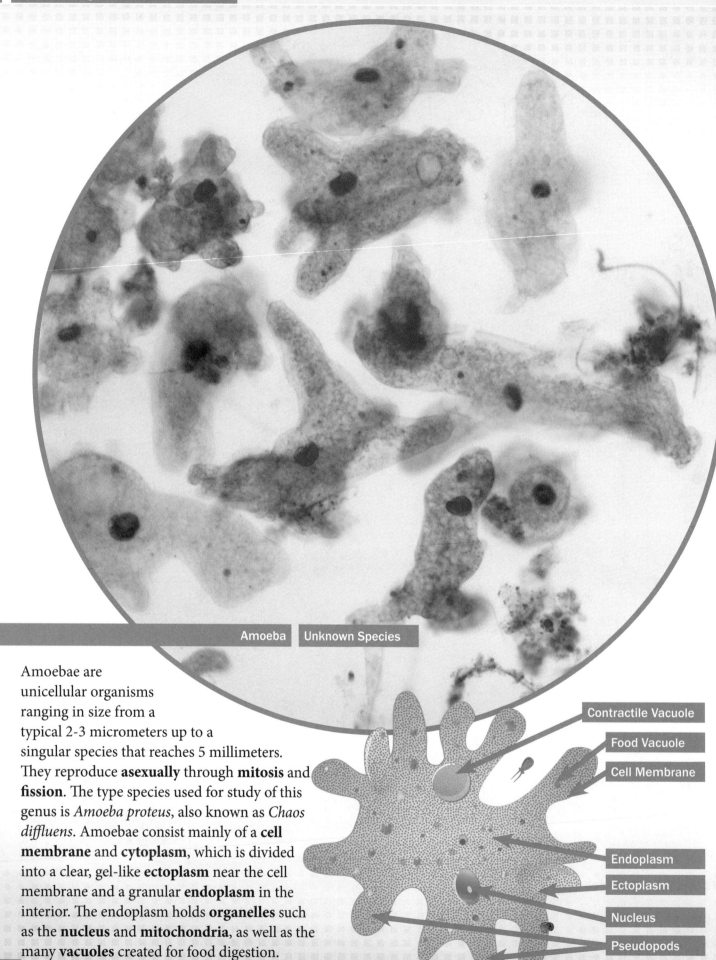

Amoeba   Unknown Species

Amoebae are unicellular organisms ranging in size from a typical 2-3 micrometers up to a singular species that reaches 5 millimeters. They reproduce **asexually** through **mitosis** and **fission**. The type species used for study of this genus is *Amoeba proteus*, also known as *Chaos diffluens*. Amoebae consist mainly of a **cell membrane** and **cytoplasm**, which is divided into a clear, gel-like **ectoplasm** near the cell membrane and a granular **endoplasm** in the interior. The endoplasm holds **organelles** such as the **nucleus** and **mitochondria**, as well as the many **vacuoles** created for food digestion.

Contractile Vacuole

Food Vacuole

Cell Membrane

Endoplasm

Ectoplasm

Nucleus

Pseudopods

Amoebae are classified as **pseudopods**, altering their shape through deliberate **protoplasmic flow** deforming their **cell membrane** into protrusions. This is how they typically feed: ingesting food particles by enveloping them in pseudopods, a process known as **phagocytosis**. Some also feed through the process of **pinocytosis**, in which dissolved nutrients that contact the cell membrane are enveloped in a small section that folds inward forming a **food vacuole**.

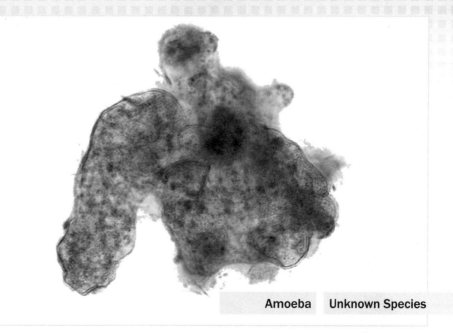

Amoeba   Unknown Species

Another type of vacuole, seen predominately in freshwater amoebae, is a **contractile vacuole**. This cavity is used to collect and periodically expel excess water that would otherwise build up through **osmotic pressure** until the amoeba burst. While amoebae have a method to deal with limited differences in salinity between their bodies and the surrounding environment, not all hazards are as easily managed.

In extremely hazardous environments, when the amoeba is threatened by toxins or a lack of moisture, many form a microbial cyst by excreting a hardened shell. In this state, amoebae become dormant until the environment is survivable again, although they can die if forced to remain **encysted** for too long. Some amoebae also form a cyst for reproduction, dividing multiple times while encysted and releasing the daughter cells simultaneously when conditions improve.

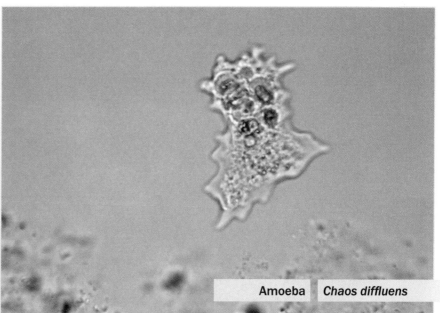

Amoeba   *Chaos diffluens*

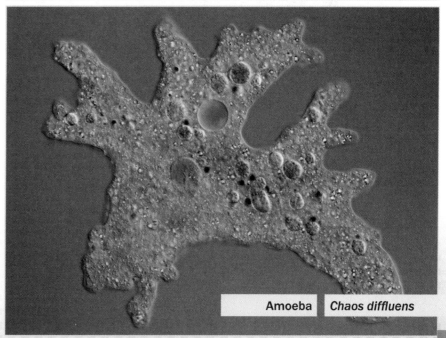

Amoeba   *Chaos diffluens*

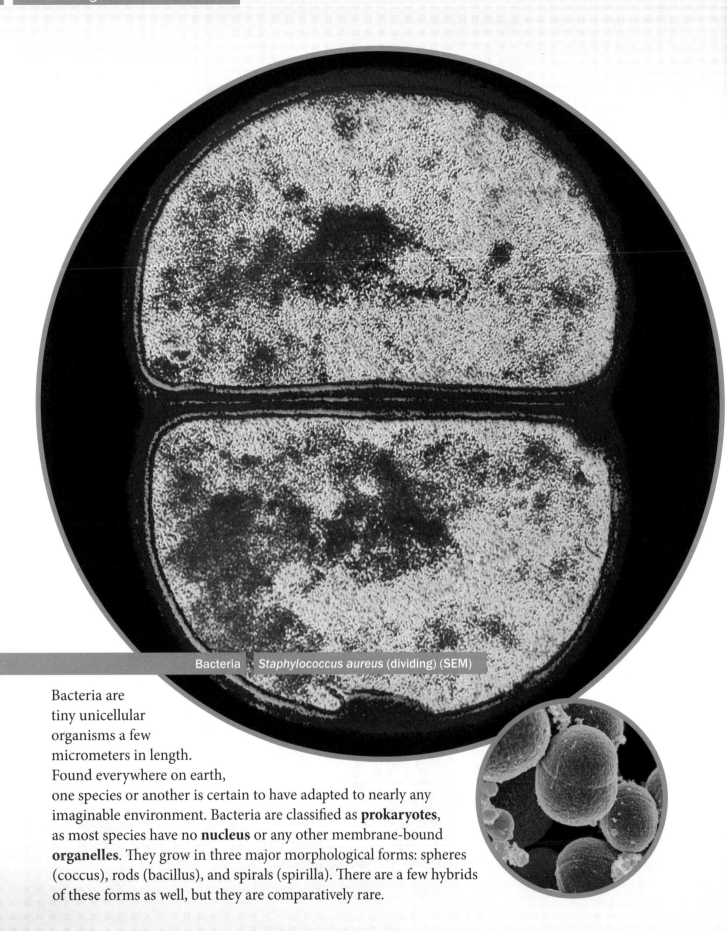

Bacteria | *Staphylococcus aureus* (dividing) (SEM)

Bacteria are tiny unicellular organisms a few micrometers in length. Found everywhere on earth, one species or another is certain to have adapted to nearly any imaginable environment. Bacteria are classified as **prokaryotes**, as most species have no **nucleus** or any other membrane-bound **organelles**. They grow in three major morphological forms: spheres (coccus), rods (bacillus), and spirals (spirilla). There are a few hybrids of these forms as well, but they are comparatively rare.

Some bacteria also have **flagella**, allowing for limited locomotion and predatory colonization. This appendage and their **cell wall** are the only distinct exterior features of their structure, while inside the cell wall bacteria have a **cell membrane** and **cytoplasm**.

One way of classifying bacteria is by the thickness of their cell wall using gram staining. Thick-walled bacteria have no outer layer of lipids, so their thicker cell wall retains a violet stain in this process. Thin-walled bacteria are known as gram-negative because they have a thin outer secondary membrane of **lipopolysaccharides** that prevents the cell wall from absorbing the violet stain, indicating resistance to some antibiotics.

Inside the cell wall, the inner cell membrane provides a diffusion barrier, regulating molecular exchange between the bacterial cytoplasm and the outside environment. Bacteria also have a cytoskeleton which helps manage the process of cell division and resource localization.

Growth and reproduction in bacteria occurs chiefly through the process of **binary fission**, where the DNA is replicated in the cell which then separates equally into two daughter cells. Some bacteria also reproduce by less symmetrical methods of division, such as **budding**, where a new organism develops from a small outgrowth on the parent.

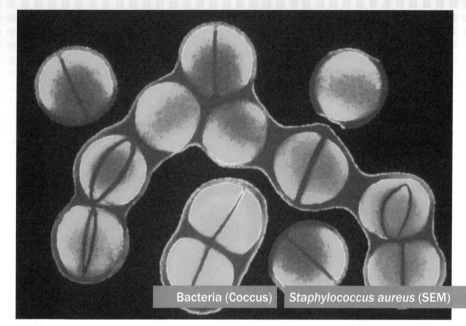

Bacteria (Coccus) *Staphylococcus aureus* (SEM)

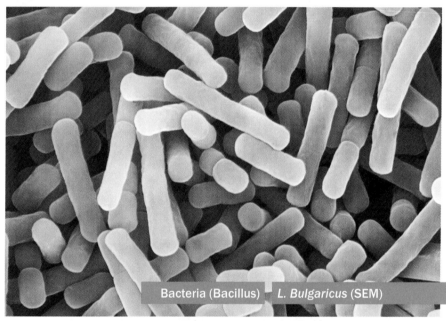

Bacteria (Bacillus) *L. Bulgaricus* (SEM)

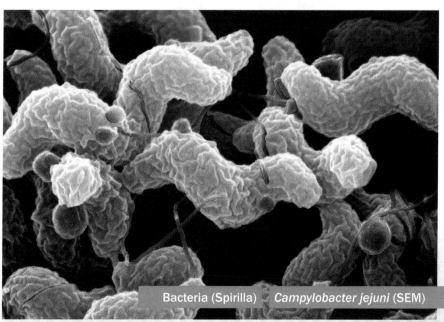

Bacteria (Spirilla) *Campylobacter jejuni* (SEM)

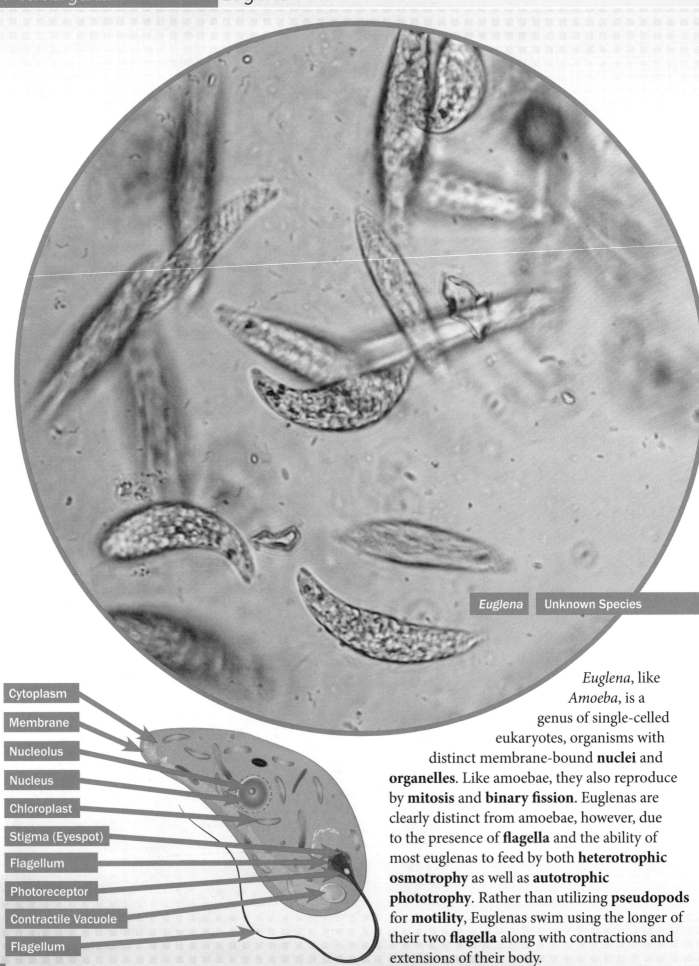

Euglena   Unknown Species

Cytoplasm

Membrane

Nucleolus

Nucleus

Chloroplast

Stigma (Eyespot)

Flagellum

Photoreceptor

Contractile Vacuole

Flagellum

Euglena, like *Amoeba*, is a genus of single-celled eukaryotes, organisms with distinct membrane-bound **nuclei** and **organelles**. Like amoebae, they also reproduce by **mitosis** and **binary fission**. Euglenas are clearly distinct from amoebae, however, due to the presence of **flagella** and the ability of most euglenas to feed by both **heterotrophic osmotrophy** as well as **autotrophic phototrophy**. Rather than utilizing **pseudopods** for **motility**, Euglenas swim using the longer of their two **flagella** along with contractions and extensions of their body.

Euglenas do not have a distinct **cell wall**, instead their exterior membrane is an elastic structure of proteins known as a pellicle, arranged in spiraling strips around the cell and supported by microtubules. This structure enables euglenas to contract and extend their bodies in order to move.

While in nutrient-rich environments, euglenas can absorb organic matter through their **cell membrane**. They also contain **chloroplasts** and use **photosynthesis** when there is sufficient light. These chloroplasts also contain pyrenoids, a form of starch energy storage which allows for survival when there is both insufficient light for photosynthesis and a lack of organic nutrients in the surrounding **medium**.

In order to be more efficient at photosynthesizing, Euglenas have a red-pigmented eyespot which filters light reaching a photosensitive **paraflagellar** body at the base of their flagella. This pigmented area does not detect light itself but blocks light from the detecting organ to varying degrees as the cell rotates (**phototaxis**), guiding the Euglena toward the light source.

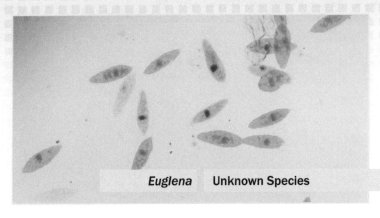

Euglena   **Unknown Species**

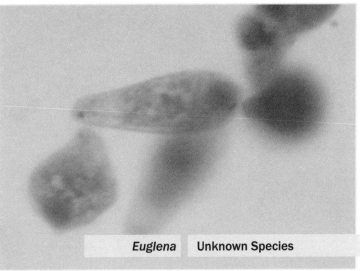

Euglena   **Unknown Species**

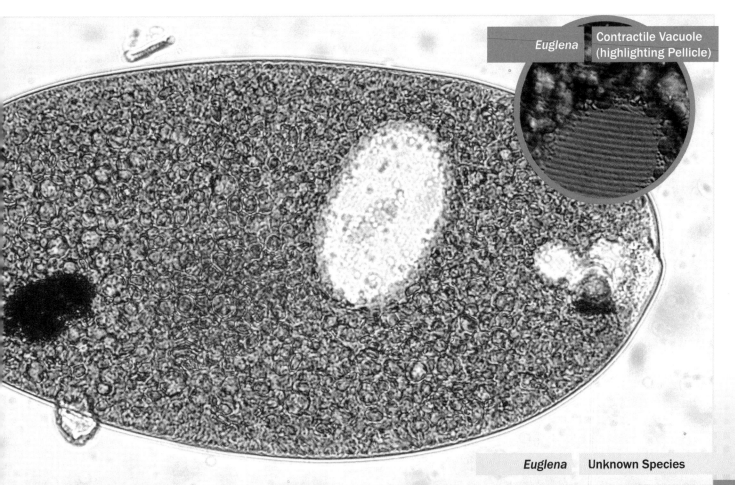

Euglena   **Contractile Vacuole (highlighting Pellicle)**

Euglena   **Unknown Species**

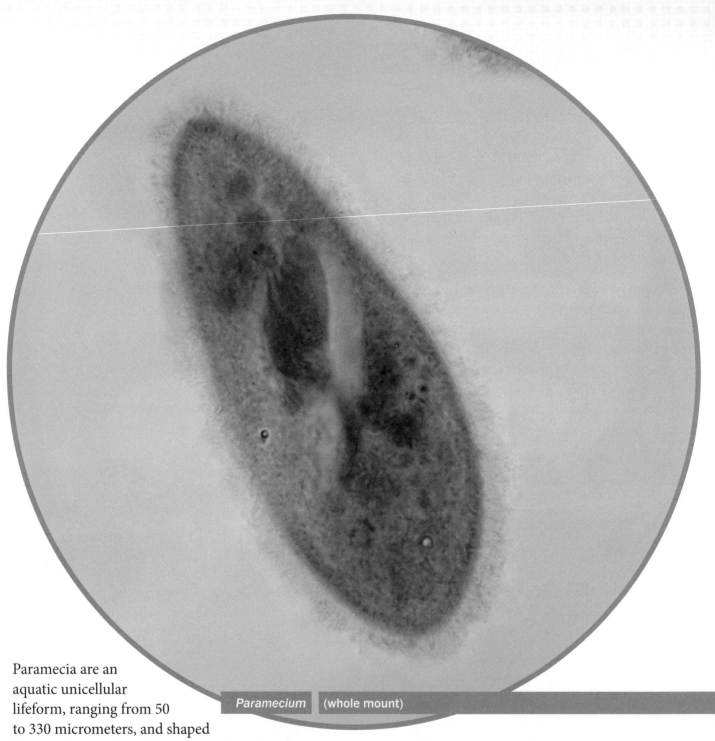

Paramecium   (whole mount)

Paramecia are an aquatic unicellular lifeform, ranging from 50 to 330 micrometers, and shaped like an elongated egg. They are uniformly covered with short, hairlike **cilia** that wave in concert to propel the paramecium through the water.

Because they live predominately in freshwater environments, paramecia have one or more **contractile vacuoles** to manage **osmotic pressure**. They also form **endosymbiotic** relationships with various other organisms which live freely in the paramecium's **cytoplasm**. These include forms of algae that supplement the paramecium's primary source of food with **photosynthetic** products, as well as bacteria known as Kappa particles that excrete compounds toxic to non-Kappa-containing paramecia, giving the host a competitive advantage.

Not all paramecia are **mixotrophs** with **algal symbionts**, but all paramecia are **heterotrophic**, consuming other small organisms. Paramecia have a deep groove, or gullet, on their anterior portion which is lined with **cilia**. These beat constantly, wafting food to the base of the gullet, where it passes through the cell mouth (cytostome) and is encapsulated into a **food vacuole** with digestive **enzymes**. These enzymes break down the ingested material while the food vacuole drifts in the paramecium's cytoplasm, shrinking as it releases the nutrients it contains. Eventually the vacuole reaches the anal pore (cytoproct), in the posterior half of the paramecium, where it ruptures and expels the remaining indigestible material.

Paramecia have a multi-nucleic structure, with a single large macronucleus controlling cellular functions and one or more micronuclei containing the genetic information necessary for reproduction, which can occur in several ways. Paramecia may reproduce **asexually** by **binary fission**, with the macronucleus and micronuclei undergoing **mitosis**. This form of reproduction does not repair any existing damage to the DNA structure, and after several hundred **iterations** will kill the cell.

Sexual reproduction may happen by conjugation between two paramecia where they join and exchange **gametes** from the division of their micronuclei by **meiosis**, or by autogamy, where a single paramecium's micronuclei divide and recombine, healing errors in each other's DNA in the process. In both of these cases, the old macronucleus is replaced by a new one generated from the healthy micronuclei, resetting the countdown to cell death.

Cilia

Pellicle

Contractile Vacuole

Radiating Canals

Micronucleus

Macronucleus

Oral Grove (Vestibulum)

Buccal Overture

Cell Mouth (Cytostome)

Food Vacuoles

Anal Pore (Cytoproct)

Cytoplasm

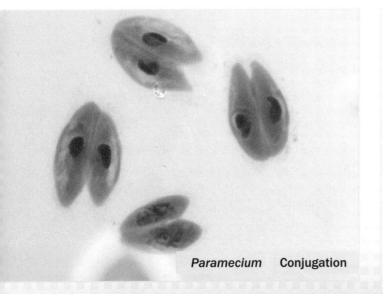

Paramecium   Conjugation

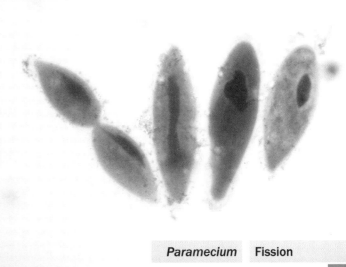

Paramecium   Fission

Microorganisms like euglenas, paramecia, and amoebae share remarkable cellular similarities: their **cytoplasmic** structure and **cellular membrane** with a **contractile vacuole** to manage **osmotic pressure**, the existence of a **nucleus** holding genetic information, and their ability to move deliberately. Even with these similarities, the three are clearly very distinct organisms, differing by their methods of movement and gaining nutrition. Bacteria are even more visibly distinct, being **prokaryotes** with no clear **organelles** at all, although they still have a cell membrane and **cytoplasm**.

Unlike the larger organisms we've seen, unicellular microorganisms like amoebae and bacteria aren't directly mentioned in the Bible, but as we've looked at them and their methods of reproduction, it's clear that the practical steps for sanitation and isolation that God gave have value in mitigating the spread of disease, especially given the available measures at the time. Numbers 19:14-16 and Leviticus 13:4-6 both give direction for ceremonial uncleanness for seven days, effectively providing clinical isolation after exposure to dead bodies or an unidentified sore in the skin.

On the positive side, God's instructions for leaving fields unharvested on the seventh year allow crucial time for bacteria and microbes in the soil and roots of the vegetation to enrich the soil with nitrates, ensuring better harvests in later years. These instructions may not have made sense at the time, but we now know that there is sound reasoning behind them.

As we study God's world of tiny wonders through a microscope, we continue to learn more and marvel at the intricacy of everything in the created world. While we use science and technology to reveal these marvelous wonders, we should be thankful not only for knowledge, but for His continued love and provision for us.

> For six years you shall sow your field, and for six years you shall prune your vineyard and gather in its fruits, but in the seventh year there shall be a Sabbath of solemn rest for the land, a Sabbath to the Lord. You shall not sow your field or prune your vineyard. You shall not reap what grows of itself in your harvest, or gather the grapes of your undressed vine. It shall be a year of solemn rest for the land. The Sabbath of the land shall provide food for you, for yourself and for your male and female slaves and for your hired worker and the sojourner who lives with you, and for your cattle and for the wild animals that are in your land: all its yield shall be for food.
>
> **Leviticus 25:3-7**

➡️ GLOSSARY

air vortices — circular patterns of rotating air

algal symbiont — algae forming a symbiotic relationship with a host organism

androecium — the group of a flower's stamens

antagonist — an organism that is naturally competitive or predatory toward another

anther — small pod at the tip of a stamen that produces pollen

antheridiophore — a gametophore (structure bearing sex organs) carrying only antheridia (male organs)

antheridium — male sex organ in non-flowering plants such as mosses and ferns, or in algae and fungi

anticoagulant — a compound that inhibits the clotting of blood

archegoniophore — a gametophore (structure bearing sex organs) carrying only archegonia (female organs)

archegonium — female sex organ in non-flowering plants such as mosses and ferns, or in algae and fungi

asexual reproduction — reproductive method where the offspring are genetic clones of a single parent, inheriting the same genetic information

autotroph — an organism that converts simple inorganic substances into organic nutrients

binary fission — a form of fission where the parent entity divides equally into two halves

budding — an asexual reproductive method where cell division occurs at a particular area on the parent's surface, creating an outgrowth which matures and then separates into a new entity

bulb scales — leaf-like layers of the bulb containing food reserves

cambium — layer of circumferential plant tissue that divides to form vascular tissues and cork, resulting in secondary growth (thickening) in plants

cell membrane — semipermeable membrane enclosing cellular cytoplasm

cellulose — insoluble polysaccharide that is the main component of cell walls and vegetable fibers

cell wall — rigid layer of polysaccharides, frequently comprised chiefly of cellulose

cercarium, cercariae — free-swimming parasitic fluke larva: period during which it transitions between hosts

chemoreceptor — chemically responsive organ or sensory cell

chlorophyll — magnesium-based green pigment that absorbs light to provide energy for photosynthesis in plants and some bacteria

chloroplast — a chlorophyll-containing plastid (small organelle) in plants where photosynthesis takes place

chromosome — a structure of nucleic acids and protein carrying genetic information found in most living cells

chrysalis — transitional state of an insect; a hard outer case enclosing an immobile pupa

cilia — short hairlike structures that work in concert to propel a microorganism, or to cause currents in surrounding fluid

collenchyma — supplementary structural cells in a plant, particularly in areas of new growth

compressive strength — ability to resist breaking when being pressed against

contractile vacuole — a specialized vacuole in some protozoans that can contract, expelling its contents from the cell

cotyledon — embryonic leaf-like structure in seed-bearing plants

cytoplasm — material inside a living cell other than the nucleus

diapause — a period of suspended development in invertebrates or embryos, particularly in unfavorable environmental conditions

differentiated — a cell that has changed from one type to another, usually more specialized, type

dioecious — plant or invertebrate animal having only either male or female sexual organs in any one individual

diploid — a cell having two complete sets of paired chromosomes (one from each parent)

disease vector — any agent conveying an infectious pathogen into another organism: may be living parasites or microbes, or inanimate matter

domesticated — cultivated and raised in a controlled environment over generations to adapt for human use

ectoplasm — outer layer of cytoplasm in amoeboid cells, resists movement

encysted — enclosed in a tough protective capsule

endoplasm — granular inner layer of cytoplasm in amoeboid cells, flows freely

endosymbiotic — one organism living inside of another one while still being mutually beneficial

enzyme — organically produced catalytic substance for facilitating biochemical reactions

fission — when a single entity divides into two or more parts, each of which regenerates to resemble the original

flagellum — slender, isolated, whiplike structure on many micro-organisms that enables swimming

flatworm — a phylum consisting of soft-bodied invertebrates such as planarians, cestodes (tapeworms), and trematodes (flukes)

food vacuole — a specialized vacuole, contains food particles and digestive enzymes

free-living — a mobile organism, not parasitic or attached to a substrate

fruiting — spore producing

gamete — a mature haploid cell of either gender capable of forming a zygote by joining another of the opposite sex

gametophyte — haploid phase in alternating generation plants that reproduces sexually into zygotes

germination — process of an organism growing from a seed

gynoecium — female part of a flower's reproductive organ, consisting of one or more pistils

haploid — a cell having a single set of unpaired chromosomes, may be reproductive cells in a diploid organism or an organism that reproduces asexually

hermaphrodite — having both male and female sexual characteristics and/or organs

heterotroph — an organism that consumes complex organic substances for nutrients

hypha — individual branching filaments composing a fungus' mycelium

hypopharynx — part of a mosquito's mouthparts that works in tandem with the labrum to create a straw-like structure

indigenous — native to or naturally occurring in an area

iteration — an individual instance of a repeating process, or the process of repeating a particular action repeatedly

labrum — structure in the upper border of the mouthparts of insects and crustaceans roughly corresponding to a lip

larva — active immature form of an insect; between egg and pupa

lignin — organic polymer that stiffens cell walls in plants

lipopolysaccharide — a large molecule consisting of a lipid (fat) and a polysaccharide like glycogen, cellulose, or a starch

mandible — any portion of the mouth or beak used to crush food; lower jaw, either portion of a bird's beak, or either of the halves of an arthropod's mouthparts used for crushing

maxilla — upper jaw in most vertebrates, and each of a pair of mouthparts used for chewing in arthropods

medium — substance in which an organism resides

megasporangium — sporangium in heterosporous plants that produces megaspores (female)

meiosis — cell division that divides the cell's genetic material, resulting in four daughter cells having half the number of chromosomes of the parent cell

microsporangium — sporangium in heterosporous plants that produces microspores (male)

miracidium, miracidia — free-swimming ciliated parasitic fluke larva, hatched from an egg

mitochondrion — organelle where respiration and energy production occur, found in large quantities in most cells

mitosis — cell division that duplicates the cell's genetic material, resulting in two daughter cells having the same number and kinds of chromosomes as the parent

mixotroph — organism that uses a variety of different energy sources, generally blending auto or phototrophy with heterotrophy

model organism — a species that is widely studied, usually as a representative of its genus due to being easy to breed and maintain in a laboratory setting. Used to help understand general biological processes.

molt, moltings — shedding of old feathers, skin, or shells to make way for new growth

monoecious — hermaphroditic plant or invertebrate animal

mosaic — a combination of diverse elements forming a coherent whole

motile — capable of independent, self-directed motion

mycelium — a network of fine white filaments binding a fungal organism together

nucleus — dense organelle containing a cell's genetic material, bound in a double membrane

ommatidium — a single one of the optical units comprising a compound eye

organelles — tiny cellular substructure that performs a specific function

osmotic pressure — a measure of how much pressure a solvent exerts through a semipermeable membrane on a solution when trying to equalize concentrations

osmotroph — organism that ingests dissolved organic compounds by osmosis for nutrition

ovule — plant structure that contains the female germ cell and develops into a seed after fertilization

ovum — female gamete (egg cell)

paraflagellar — adjacent to or part of the flagella

parasitic, parasite — a relationship between organisms where one relies on the other one, typically without providing a reciprocal benefit, the dependent organism in such a relationship

parenchyma — functional, or metabolic, tissues of an organism (originally used for animal tissues, now also used for plants)

perennial — a plant that lives for multiple years

phagocytosis — ingestion of material by active envelopment and extending the cell around the material

pheromone — chemical substance released by an individual organism which is used for communication, affecting the behavior and physiology of other members of the same species

phloem — vascular plant tissue carrying photosynthesized products from the leaves downward

photoautotroph — an organism that uses sunlight to generate its own food from organic molecules

photoreceptor — structure or sensory cell(s) in an organism that respond to light

photosynthesis — process of using light to synthesize foods from carbon dioxide and water, generating oxygen as a byproduct

phototaxis — motile movement in response to light

phototroph — an organism that uses sunlight to synthesize organic nutrients

pinocytosis — ingestion of material by budding a small vesicle into the cell after the material comes into contact with it

pistil — individual portion of a flower's gynoecium, comprised of an ovary at the base, an elongated style, and a structure for capturing pollen at the top called a stigma

polysaccharide — several sugar molecules bonded together forming a carbohydrate

proboscis — an elongated mouthpart, typically tubular, used for suction

prokaryotes — a classification of unicellular organism with the characteristic of having no distinct nucleus or specialized membrane-bound organelles

protoplasmic flow — the movement of cytoplasm inside the cell, used for motion and nutrient dissemination in amoebae and some other large cells

pseudopods — temporary extension from the surface of an amoeboid cell, used for movement or feeding

pupa — inactive transitional form of an immature insect; between larval and adult forms

pupate — process of forming a pupa and chrysalis

raptorial — predatory, used for seizing prey

redia, rediae — initial larva form of a parasitic fluke; forms cercaria. Some flukes bypass this stage, growing cercariae immediately from sporocysts.

respiration — the process of energy production in living organisms, typically through the oxidation of organic substances

rhizoid — simple structure resembling roots in plants lacking true roots

rhizome — horizontal underground stem that shoots off periodic lateral stems and roots

scales — protective cells covering areas of a thallus that will develop into male or female sexual organs

sclerenchyma — main structural cells in a plant, typically woody, with very thick cell walls

somatic — related to the body in general, as opposed to germ cells specifically related to reproduction

sporangium — receptacle containing spores in asexual reproduction of plants

sporocyst — initial stage of infection by a parasitic fluke in an invertebrate host, developed from a miracidium

sporophyte — diploid phase in alternating generation plants, reproduces asexually into spores

stamen — a flower's male fertilizing organ, comprised of a stalk-like filament and pollen-containing anther

stigma — structure at the top of a pistil that captures pollen, also another word for eyespot

striated — marked by transverse (crossways) dark and light bands

style — tubelike structure between the stigma and ovary of a flower, either draws a pollen grain down or allows a pollen tube to grow to reach the ovule

substrate — surface or material that an organism is attached to or grows from, may also provide nourishment

symbiotic — a relationship between organisms living in close physical proximity to their mutual benefit

tensile strength — ability to resist breaking when being pulled apart

vacuole — small cavity or space in a cell, typically fluid-filled and enclosed by a membrane

venter — flask-shaped structure in liverworts containing an egg cell

xylem — vascular woody plant tissue carrying water and nutrients upward from the root

zygote — diploid cell formed after two haploid gametes join (a fertilized ovum)

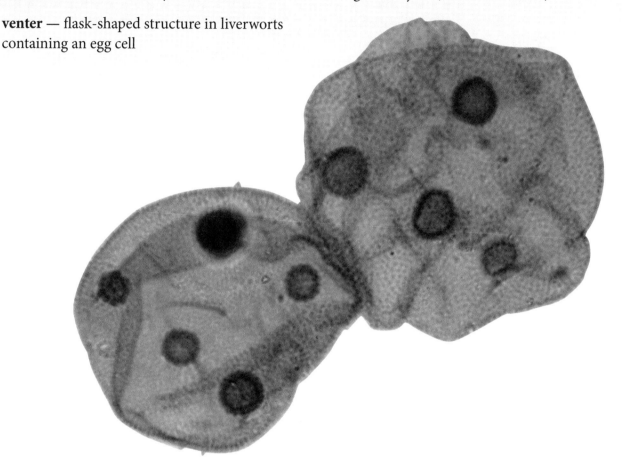

PHOTO CREDITS:

➡ INDEX

ORGANISM / GENUS

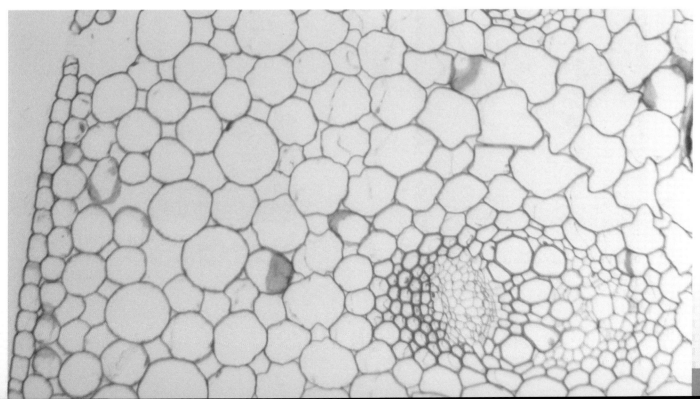

CREATION BASED
HIGH SCHOOL SCIENCE

WITH LABS

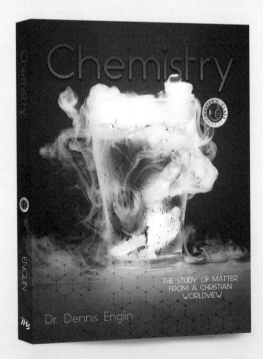

MASTER'S CLASS: CHEMISTRY
GRADE 10-12 | 978-1-68344-134-2

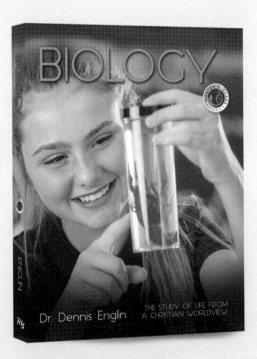

MASTER'S CLASS: BIOLOGY
GRADE 9-12 | 978-1-68344-152-6

MASTERBOOKS.COM
— *Where Faith Grows!* —

TO SEE OUR FULL LINE
OF FAITH-BUILDING
CURRICULUM